AF560285

Fruit and Vegetable Processing

Fruit and Vegetable Processing

Vikram Sharma

RANDOM PUBLICATIONS
NEW DELHI (INDIA)

Fruit and Vegetable Processing

ISBN 978-93-5111-521-2

Published in 2015 in India by

RANDOM PUBLICATIONS

4376-A/4B, Gali Murari Lal, Ansari Road
New Delhi-110 002

Phone : +9111-43580356, 011-23289044, 011-43142548
e-mail: sales@randompublications.com,
info@randompublications.com, randomexports@gmail.com

Reprinted 2021

Type Setting by : Friends Media, Delhi-110089
Digitally Printed at : Replika Press Pvt. Ltd.

Preface

Fruit and vegetable production and consumption in the world have shown a marked upward trend over the past several years. Rising consumer demand has come with greater awareness of food safety issues and increased need for convenience and quality. The horticultural sector has responded to these demands, and is currently at different stages of development across the world.

Historically speaking, processing of fruits and vegetables in the simplest form like pickling, sun-drying and/or making preserves has been practised from very ancient times almost in every home. Since then it has been progressing fairly well and has been meeting the entire local demand and has successfully entered the exports market. Today, fruits and vegetables are processed into a variety of products such as juices and concentrates, pulp, canned and dehydrated products, jams and jellies, pickles and chutneys etc.

The extent of processing of fruits and vegetables varies from one country to another. Assuring the safety and quality of the produce necessitates the selection of high quality horticultural produce for processing, and the implementation of good practice during processing operations in order to maintain produce quality and assure safety of the final product.

This book reviews in detail from a theoretical and practical perspective, the critical issues that must be addressed if the processed fruit and vegetable products are to meet consumer and market demand for convenience, quality and safety. Written in a simple, easy-to-read format, it will be of practical value, to small processors, trainers, extension workers and non-governmental organizations who provide training and support to individuals engaged in fruit and vegetable processing for sale. It is also provides a useful source of information for consumers of processed horticultural products.

Author

Contents

1

Introduction to Fruit and Vegetable Processing

In developing countries agriculture is the mainstay of the economy. As such, it should be no surprise that agricultural industries and related activities can account for a considerable proportion of their output. Of the various types of activities that can be termed as agriculturally based, fruit and vegetable processing are among the most important.

Both established and planned fruit and vegetable processing projects aim at solving a very clearly identified development problem. This is that due to insufficient demand, weak infrastructure, poor transportation and perishable nature of the crops, the grower sustains substantial losses. During the post-harvest glut, the loss is considerable and often some of the produce has to be fed to animals or allowed to rot.

Even established fruit and vegetable canning factories or small/medium scale processing centres suffer huge loss due to erratic supplies. The grower may like to sell his produce in the open market directly to the consumer, or the produce may not be of high enough quality to process even though it might be good enough for the table. This means that processing capacities will be seriously underexploited.

The main objective of fruit and vegetable processing is to supply wholesome, safe, nutritious and acceptable food to consumers throughout the year. Fruit and vegetable processing projects also aim to replace imported products like squash, yams, tomato sauces, pickles, etc., besides earning foreign exchange by exporting finished or semi-processed products.

The fruit and vegetable processing activities have been set up, or have to be established in developing countries for one or other of the following reasons:

— diversification of the economy, in order to reduce present dependence on one export commodity;
— government industrialisation policy;
— reduction of imports and meeting export demands;
— stimulate agricultural production by obtaining marketable products;
— generate both rural and urban employment;
— reduce fruit and vegetable losses;
— improve farmers' nutrition by allowing them to consume their own processed fruit and vegetables during the off-season;
— generate new sources of income for farmers/artisans;
— develop new value-added products

What Fruit and Vegetables can be Processed?

Practically any fruit and vegetable can be processed, but some important factors which determine whether it is worthwhile are:

a. the demand for a particular fruit or vegetable in the processed form;
b. the quality of the raw material, i.e. whether it can withstand processing;
c. regular supplies of the raw material.

For example, a particular variety of fruit which may be excellent to eat fresh is not necessarily good for processing. Processing requires frequent handling, high temperature and pressure.

Many of the ordinary table varieties of tomatoes, for instance, are not suitable for making paste or other processed products. A particular mango or pineapple may be very tasty eaten fresh, but when it goes to the processing centre it may fail to stand up to the processing requirements due to variations in its quality, size, maturity, variety and so on.

Even when a variety can be processed, it is not suitable unless large and regular supplies are made available. An important processing centre or a factory cannot be planned just to rely on seasonal gluts; although it can take care of the gluts it will not run economically unless regular supplies are guaranteed.

To operate a fruit and vegetable processing centre efficiently it is of utmost importance to pre-organise growth, collection and transport of suitable raw material, either on the nucleus farm basis or using outgrowers.

Processing Planning

The secret of a well planned fruit and vegetable processing centre is that it must be designed to operate for as many months of the year as possible. This means the facilities, the buildings, the material handling and the equipment itself must be inter-linked and coordinated properly to allow as many products as possible to be handled at the same time, and yet the equipment must be versatile enough to be able to handle many products without major alterations.

A typical processing centre or factory should process four or five types of fruits harvested at different times of the year and two or three vegetables. This processing unit must also be capable of handling dried/dehydrated finished products, juices, pickles, tomato juice, ketchup and paste, jams, jellies and marmalades, semi-processed fruit products.

Advanced planning is necessary to process a large range of products in varied weather and temperature conditions, each requiring a special set of manufacturing and packaging formulae. The end result of the efforts should be a well-managed processing unit with lower initial investment.

A unit which is sensibly laid out and where one requirement co-relates to another, with a sound costing analysis, leads to an integrated operation.

Instead of over-sophisticated machinery, a sensible simple processing unit may be required when planned production is not very large and is geared mainly to meet the demand of the domestic market.

Location

The basic objective is to choose the location which minimises the average production cost, including transport and handling.

It is an advantage, all other things being equal, to locate a processing unit near the fresh raw material supply. It is a necessity for proper handling of the perishable raw materials, it allows the processing unit to allow the product to reach its best stage of maturation and lessens injury from handling and deterioration from changes during long transportation after harvesting.

An adequate supply of good water, availability of manpower, proximity to rail or road transport facilities and adequate markets are other important requirements.

Processing Systems

Small-Scale Processing

This is done by small-scale farmers for personal subsistence or for sale in nearby markets. In this system, processing requires little investment: however, it is time consuming and tedious. Until recently, small-scale processing satisfied the needs of rural and urban populations. However, with the rising rates of population and urbanisation growth and their more diversified food demands, there is need for more processed and diversified types of food.

Intermediate-Scale Processing

In this scale of processing, a group of small-scale processors pool their resources. This can also be done by individuals. Processing is based on the technology used by small-scale processors with differences in the type and capacity of equipment used. The raw materials are usually grown by the processors themselves or are purchased on contract from other farmers. These operations are usually located on the production site of in order to assure raw materials availability and reduce cost of transport. This system of processing can provide quantities of processed products to urban areas.

Large-Scale Processing

Processing in this system is highly mechanised and requires a substantial supply of raw materials for economical operation. This system requires a large capital investment and high technical and managerial skills. Because of the high demand for foods in recent years many large-scale factories were established in developing countries. Some succeeded, but the majority failed, especially in West Africa. Most of the failures were related to high labour inputs and relatively high cost, lack of managerial skills, high cost and supply instability of raw materials and changing governmental policies. Perhaps the most important reason for failure was lack of adequate quantity and regularity of raw material supply to factories. Despite the failure of these commercial operations, they should be able to succeed with better planning and management, along with the undertaking of more in-depth feasibility studies.

It can be concluded that all three types of processing systems have a place in developing countries to complement crop production to meet food demand. Historically, however, small and intermediate scale processing proved to be more successful than large-scale processing in developing countries.

Choice of Processing Technologies for Developing Countries

FAO maintains, that the basis for choosing a processing technology for developing countries ought to be to combine labour, material resources and capital so that not only the type and quantity of goods and services produced are taken into account, but also the distribution of their benefits and the prospects of overall growth. These should include:

a. increasing farmer/artisan income by the full utilisation of available indigenous raw material and local manufacturing of part or all processing equipment;

b. cutting production costs by better utilisation of local natural resources (solar energy) and reducing transport costs;

c. generating and distributing income by decentralising processing activities and involving different beneficiaries in processing activities (investors, newly employed, farmers and small-scale industry);

d. maximising national output by reducing capital expenditure and royalty payments, more effectively developing balance-of-payments deficits through minimising imports (equipment, packing material, additives), and maximising export-oriented production;

e. maximising availability of consumer goods by maximisation of high-quality, standard processed produce for internal and export markets, reducing post-harvest losses, giving added value to indigenous crops and increasing the volume and quality of agricultural output.

Knowledge and control of the means of production, local manufacturing of processing equipment and development of appropriate/new technologies and more suitable raw material for processing must all be better researched.

Decentralisation of activities must be maintained and coordinated. The introduction of more sophisticated processing equipment and packaging material must be subordinated to internal and export marketing references.

Choosing a technology solely to maximise profits can actually work against true development. Choice should also be based on a solid, long-term market opportunity to ensure viability.

General Properties of Fruit and Vegetables

Fruit and vegetables have many similarities with respect to their compositions, methods of cultivation and harvesting, storage properties and processing. In fact, many vegetables may be considered fruit in the true botanical sense. Botanically, fruits are those portions of the plant which house seeds. Therefore such items as tomatoes, cucumbers, eggplant, peppers, and others would be classified as fruits on this basis.

However, the important distinction between fruit and vegetables has come to be made on an usage basis. Those plant items that are generally eaten with the main course of a meal are considered to be vegetables. Those that are commonly eaten as dessert are considered fruits.

Vegetables are derived from various parts of plants and it is sometimes useful to associate different vegetables with the parts of the plant they represent since this provides clues to some of the characteristics we may expect in these items. A classification of vegetables based on morphological features is seen in Table 1.

Table 1. Classification of Vegetables

Category	*Examples*
Earth vegetables roots	sweet potatoes, carrots
modified stems tubers	potatoes
modified buds bulbs	onions, garlic
Herbage vegetables	
leaves	cabbage, spinach, lettuce
petioles (leaf stalk)	celery, rhubarb
flower buds	cauliflower, artichokes
sprouts, shoots (young stems)	asparagus, bamboo shoots
Fruit vegetables	
legumes	peas, green beans
cereals	sweet corn
vine fruits	squash, cucumber
berry fruits	tomato, egg plant
tree fruits	avocado, breadfruit

Fruit as a dessert item, is the mature ovaries of plants with their seeds. The edible portion of most fruit is the fleshy part of the pericarp or vessel surrounding the seeds. Fruit in general is acidic and sugary. They commonly are grouped into several major divisions, depending principally upon botanical structure, chemical composition and climatic requirements.

Berries are fruit which are generally small and quite fragile. Grapes are also physically fragile and grow in clusters. Melons, on the other hand, are large and have a tough outer rind. Drupes (stone fruit) contain single pits and include such items as apricots, cherries, peaches and plums. Pomes contain many pits, and are represented by apples, quince and pears.

Citrus fruit like oranges, grapefruit and lemons are high in citric acid. Tropical and subtropical fruits include bananas, dates, figs, pineapples, mangoes, and others which require warm climates, but exclude the separate group of citrus fruits.

The compositions of representative vegetables and fruits in comparison with a few of the cereal grains are seen in Table 2.

Tabel 2. Typical percentage composition of foods from plant origin Percentage Composition- Edible Portion

Food	*Carbo-hydrate*	*Protein*	*Fat*	*Ash*	*Water*
Cereals					
wheat flour, white	73.9	10.5	1.9	1.7	12
rice, milled, white	78.9	6.7	0.7	0.7	13
maize, whole grain	72.9	9.5	4.3	1.3	12
Earth vegetables					
potatoes, white	18.9	2.0	0.1	1.0	78
sweet potatoes	27.3	1.3	0.4	1.0	70
Vegetables					
carrots	9.1	1.1	0.2	1.0	88.6
radishes	4.2	1.1	0.1	0.9	93.7
asparagus	4.1	2.1	0.2	0.7	92.9
beans, snap, green	7.6	2.4	0.2	0.7	89.1
peas, fresh	17.0	6.7	0.4	0.9	75.0
lettuce	2.8	1.3	0.2	0.9	94.8
Fruit					
banana	24.0	1.3	0.4	0.8	73.5
orange	11.3	0.9	0.2	0.5	87.1
apple	15.0	0.3	0.4	0.3	84.0
strawberries	8.3	0.8	0.5	0.5	89.9

Compositions of vegetables and fruit not only vary for a given kind in according to botanical variety, cultivation practices, and weather, but change with the degree of maturity prior to harvest, and the condition of ripeness, which is progressive after harvest and is further influenced by storage conditions. Nevertheless, some generalisations can be made.

Most fresh vegetables and fruit are high in water content, low in protein, and low in fat. In these cases water contents will generally be greater than 70% and frequently greater than 85%.

Commonly protein content will not be greater than 3.5% or fat content greater than 0.5 %. Exceptions exist in the case of dates and raisins which are substantially lower in moisture but cannot be considered fresh in the same sense as other fruit. Legumes such as peas and certain beans are higher in protein; a few vegetables such as sweet corn which are slightly higher in fat and avocados which are substantially higher in fat.

Vegetables and fruit are important sources of both digestible and indigestible carbohydrates. The digestible carbohydrates are present largely in the form of sugars and starches while indigestible cellulose provides roughage which is important to normal digestion.

Fruit and vegetables are also important sources of minerals and certain vitamins, especially vitamins A and C. The precursors of vitamin A, including beta-carotene and certain other carotenoids, are to be found particularly in the yellow-orange fruit and vegetables and in the green leafy vegetables.

Citrus fruit are excellent sources of vitamin C, as are green leafy vegetables and tomatoes. Potatoes also provide an important source of vitamin C for the diets of many countries. This is not so much due to the level of vitamin C in potatoes which is not especially high but rather to the large quantities of potatoes consumed.

Chemical composition

Water

Vegetal cells contain important quantities of water. Water plays a vital role in the evolution and reproduction cycle and in physiological processes. It has effects on the storage period length and on the consumption of tissue reserve substances.

In vegetal cells, water is present in following forms:

— bound water or dilution water which is present in the cell and forms true solutions with mineral or organic substances;

— colloidal bound water which is present in the membrane, cytoplasm and nucleus and acts as a swelling agent for these colloidal structure substances; it is very difficult to remove during drying/dehydration processes;

— constitution water, directly bound on the chemical component molecules and which is also removed with difficulty.

Vegetables contain generally 90-96% water while for fruit normal water content is between 80 and 90%.

Mineral substances

Mineral substances are present as salts of organic or inorganic acids or as complex organic combinations (chlorophyll, lecithin, etc.); they are in many cases dissolved in cellular juice.

Vegetables are more rich in mineral substances as compared with fruits. The mineral substance content is normally between 0.60 and 1.80% and more than 60 elements are present; the major elements are: K, Na, Ca, Mg, Fe, Mn, Al, P. Cl, S.

Among the vegetables which are especially rich in mineral substances are: spinach, carrots, cabbage and tomatoes. Mineral rich fruit includes: strawberries, cherries, peaches and raspberries. Important quantities of potassium (K) and absence of sodium chloride (NaCl) give a high dietetic value to fruit and to their processed products. Phosphorus is supplied mainly by vegetables.

Vegetables usually contain more calcium than fruit; green beans, cabbage, onions and beans contain more than 0.1% calcium. The calcium/ phosphorus or Ca/P ratio is essential for calcium fixation in the human body; this value is considered normal at 0.7 for adults and at 1.0 for children. Some fruit are important for their Ca/P ratio above 1.0: pears, lemons, oranges and some temperate climate mountain fruits and wild berries.

Even if its content in the human body is very low, iron (Fe) has an important role as a constituent of haemoglobin. Main iron sources are apples and spinach.

Salts from fruit have a basic reaction; for this reason fruit consumption facilitates the neutralisation of noxious uric acid reactions and contributes to the acid-basic equilibrium in the blood.

Carbohydrates

Carbohydrates are the main component of fruit and vegetables and represent more than 90% of their dry matter. From an energy point of view carbohydrates represent the most valuable of the food components; daily adult intake should contain about 500 g carbohydrates.

Carbohydrates play a major role in biological systems and in foods. They are produced by the process of photosynthesis in green plants. They may serve as structural components as in the case of cellulose; they may be stored as energy reserves as in the case of starch in plants; they may function as essential components of nucleic acids as in the case of ribose; and as components of vitamins such as ribose and riboflavin.

Carbohydrates can be oxidised to furnish energy, and glucose in the blood is a ready source of energy for the human body. Fermentation of carbohydrates by yeast and other microorganisms can yield carbon dioxide, alcohol, organic acids and other compounds.

Some properties of sugars. Sugars such as glucose, fructose, maltose and sucrose all share the following characteristics in varying degrees, related to fruit and vegetable technology:

— they supply energy for nutrition;
— they are readily fermented by micro-organisms;
— in high concentrations they prevent the growth of micro-organisms, so they may be used as a preservative;
— on heating they darken in colour or caramelise;
— some of them combine with proteins to give dark colours known as the browning reaction.

Some properties of starches:

— They provide a reserve energy source in plants and supply energy in nutrition;
— they occur in seeds and tubers as characteristic starch granules.

Some properties of celluloses and hemicelluloses:

— They are abundant in the plant kingdom and act primarily as supporting structures in the plant tissues;

— they are insoluble in cold and hot water;

— they are not digested by man and so do not yield energy for nutrition;

— the fibre in food which produces necessary roughage is largely cellulose.

Some properties of pectins and carbohydrate gums.

— Pectins are common in fruits and vegetables and are gum-like (they are found in and between cell walls) and help hold the plant cells together;

— pectins in colloidal solution contribute to viscosity of the tomato paste;

— pectins in solution form gels when sugar and acid are added; this is the basis of jelly manufacture.

Fats

Generally fruit and vegetables contain very low level of fats, below 0.5%. However, significant quantities are found in nuts (55%), apricot kernel (40%), grapes seeds (16%), apple seeds (20%) and tomato seeds (18%).

Organic Acids

Fruit contains natural acids, such as citric acid in oranges and lemons, malic acid of apples, and tartaric acid of grapes. These acids give the fruits tartness and slow down bacterial spoilage.

We deliberately ferment some foods with desirable bacteria to produce acids and this give the food flavour and keeping quality. Examples are fermentation of cabbage to produce lactic acid and yield sauerkraut and fermentation of apple juice to produce first alcohol and then acetic acid to obtain vinegar.

Organic acids influence the colour of foods since many plant pigments are natural pH indicators.

With respect to bacterial spoilage, a most important contribution of organic acids is in lowering a food's pH. Under anaerobic conditions and slightly above a pH of 4.6, Clostridium botulinum can grow and produce lethal toxins. This hazard is absent from foods high in organic acids resulting in a pH of 4.6 and less.

Acidity and sugars are two main elements which determine the taste of fruit. The sugar/acid ratio is very often used in order to give a technological characterisation of fruits and of some vegetables.

Nitrogen-containing Substances

These substances are found in plants as different combinations: proteins, amino acids, amides, amines, nitrates, etc. Vegetables contain between 1.0 and 5.5 % while in fruit nitrogen-containing substances are less than 1% in most cases.

Among nitrogen containing substances the most important are proteins; they have a colloidal structure and, by heating, their water solution above 50°C an one-way reaction makes them insoluble. This behaviour has to be taken into account in heat processing of fruits and vegetables.

From a biological point of view vegetal proteins are less valuable then animal ones because in their composition all essential amino-acids are not present.

Vitamins

Vitamins are defined as organic materials which must be supplied to the human body in small amounts apart from the essential amino-acids or fatty acids.

Vitamins function as enzyme systems which facilitate the metabolism of proteins, carbohydrates and fats but there is growing evidence that their roles in maintaining health may extend yet further.

The vitamins are conveniently divided into two major groups, those that are fat-soluble and those that are water-soluble. Fat-soluble vitamins are A, D, E and K. Their absorption by the body depends upon the normal absorption of fat from the diet. Water-soluble vitamins include vitamin C and several members of the vitamin B complex.

Vitamin A or Retinol.

This vitamin is found as such only in animal materials - meat, milk, eggs and the like. Plants contain no vitamin A but contain its precursor, beta-carotene. Man needs either vitamin A or beta-carotene which he can easily convert to vitamin A. Beta-carotene is found in the orange and yellow vegetables as well as the green leafy vegetables, mainly carrots, squash, sweet potatoes, spinach and kale.

A deficiency of vitamin A leads to night blindness, failure of normal bone and tooth development in the young and diseases of epithelial cells and membrane of the nose, throat and eyes which decrease the body's resistance to infection.

Vitamin C

Vitamin C is the anti-scurvy vitamin. Lack of it causes fragile capillary walls, easy bleeding of the gums, loosening of teeth and bone joint diseases. It is necessary for the normal formation of the protein collagen, which is an important constituent of skin and connective tissue. Like vitamin E, vitamin C favours the absorption of iron.

Vitamin C, also known as ascorbic acid, is easily destroyed by oxidation especially at high temperatures and is the vitamin most easily lost during processing, storage and cooking. Excellent sources of vitamin C are citrus fruits, tomatoes, cabbage and green peppers. Potatoes also are a fair source (although the content of vitamin C is relatively low) because we consume large quantities of potatoes.

Enzymes

Enzymes are biological catalysts that promote most of the biochemical reactions which occur in vegetable cells.

Some properties of enzymes important in fruit and vegetable technology are the following:

— in living fruit and vegetables enzymes control the reactions associated with ripening;

— after harvest, unless destroyed by heat, chemicals or some other means, enzymes continue the ripening process, in many cases to the point of spoilage - such as soft melons or overripe bananas;

— because enzymes enter into a vast number of biochemical reactions in fruits and vegetable, they may be responsible for changes in flavour, colour, texture and nutritional properties;

— the heating processes in fruit and vegetables manufacturing/processing are designed not only to destroy micro-organisms but also to deactivate enzymes and so improve the fruit and vegetables' storage stability.

Enzymes have an optimal temperature - around +50°C where their activity is at maximum. Heating beyond this optimal temperature deactivates the enzyme. Activity of each enzyme is also characterised by an optimal pH.

In fruit and vegetable storage and processing the most important roles are played by the enzymes classes of hydrolases (lipase, invertase, tannase, chlorophylase, amylase, cellulase) and oxidoreductases (peroxidase, tyrosinase, catalase, ascorbinase, polyphenoloxidase).

Turgidity and Texture

The range of textures that are encountered in fresh and cooked vegetables and fruit is indeed great, and to a large extent can be explained in terms of changes in specific cellular components. Since plants tissues generally contain more than two-thirds water, the relationships between these components and water further determine textural differences.

Cell Turgidity. Quite apart from other contributing factors, the state of turgidity, determined by osmotic forces, plays a paramount role in the texture of fruit and vegetables. The cell walls of plant tissues have varying degrees of elasticity and are largely permeable to water and ions as well as to small molecules.

The membranes of the living protoplast are semi-permeable, that is they allow passage of water but are selective with respect to transfer of dissolved and suspended materials.

The cell vacuoles contain most of the water in plant cells and sugars, acids, salts, amino acids, some water-soluble pigments and vitamins, and other low molecular weight constituents are dissolved in this water.

In the living plant, water taken up by the roots passes through the cell walls and membranes into the cytoplasm of the protoplasts and into the vacuoles to establish a state of osmotic equilibrium within the cells.

The osmotic pressure within the cell vacuoles and within the protoplasts pushes the protoplasts against the cell walls and causes them to stretch slightly in accordance with their elastic properties. This is the situation in the growing plant and the harvested live fruit or vegetable which is responsible for desired plumpness, succulence, and much of the crispness.

When plant tissues are damaged or killed by storage, freezing, cooking, or other causes, an important major change that results is denaturation of the proteins of cell membranes resulting in the loss of perm-selectivity. Without perm-selectivity the state of osmotic pressure in cell vacuoles and protoplasts cannot exist, and water and dissolved substances are free to diffuse out of the cells and leave the remaining tissue in a soft and wilted condition.

Other Factors Affecting Texture. The existence of a high degree of turgidity in live fruit and vegetables or whether a relative state of flabbiness develops from loss of osmotic pressure as well as final texture depends on several cell constituents.

Cellulose, Hemicellulose, and Lignin. Cell walls in young plants are very thin and are composed largely of cellulose. As the plant ages cell walls tend to thicken and become higher in hemicellulose and in lignin. These materials are fibrous and tough and are not significantly softened by cooking.

Pectic Substances. The complex polymers of sugar acid derivatives include pectin and closely related substances. The cement-like substance found especially in the middle lamella which helps hold plant cells to one another is a water-insoluble pectic substance.

On mild hydrolysis it yields water-soluble pectin which can form gels or viscous colloidal suspensions with sugar and acid. Certain water-soluble pectic substances also react with metal ions, particularly calcium, to form water-insoluble salts such as calcium pectates. The various pectic substances may influence texture of vegetables and fruits in several ways.

When vegetables or fruit are cooked, some of the water-insoluble pectic substance is hydrolysed into water-soluble pectin. This results in a degree of cell separation in the tissues and contributes to tenderness. Since many fruits and vegetables are somewhat acidic and contain sugars the soluble pectin also tends to form colloidal suspensions which will thicken the juice or pulp of these products.

Fruit and vegetables also contain a natural enzyme which can further hydrolyse pectin to the point where the pectin loses much of its gel forming property. This enzyme is known as pectin methyl esterase. Materials such as tomato juice or tomato paste will contain both pectin and pectin methyl esterase.

If freshly prepared tomato juice or paste is allowed to stand the original viscosity gradually decreases due to the action of pectin methyl esterase on pectin gel.

This can be prevented if the tomato products are quickly heated to a temperature of about 82°C (180 F°) to deactivate the pectin methyl esterase liberated from broken cells before it has a chance to hydrolyse the pectin. Such a treatment is commonly practiced in the manufacture of tomato juice products. This is known as the "hot-break process" and yields products of high viscosity.

In contrast, where low viscosity products are desired no heat is used and enzyme activity is allowed to proceed. This is "cold-break" process. After sufficient decrease in viscosity is achieved the product can be heat treated, as in canning, to preserve it for long term storage.

It is often also desirable to firm the texture of fruit and vegetables, especially when products are normally softened by processing. In this case advantage is taken of the reaction between soluble pectic substances and calcium ions which form calcium pectates. These calcium pectates are water insoluble and when they are produced within the tissues of fruit and vegetables they increase structural rigidity. Thus, it is common commercial practice to add low levels of calcium salts to tomatoes, apples, and other vegetables and fruits prior to canning or freezing.

Sources of Colour and Colour Changes

In addition to a great range of textures, much of the interest that fruits and vegetables add to our diets is due to their delightful and variable colours. The pigments and colour precursors of fruit and vegetables occur for the most part in the cellular plastic inclusions such as the chloroplasts and other chromoplasts, and to a lesser extent dissolved in fat droplets or water within the cell protoplast and vacuoles.

These pigments are classified into four major groups which include the chlorophylls, carotenoids, anthocyanins, and anthoanthins. Pigments belonging to the latter two groups also are referred to as flavonoids, and include the tannins.

The Chlorophylls. The chlorophylls are contained mainly within the chloroplasts and have a primary role in the photosynthetic production of carbohydrates from carbon dioxide and water. The bright green colour of leaves and other parts of plants is largely due to the oilsoluble chlorophylls, which in nature are bound to protein molecules in highly organised complexes.

When the plant cells are killed by ageing, processing, or cooking, the protein of these complexes is denatured and the chlorophyll may be released. Such chlorophyll is highly unstable and rapidly changes in colour to olive green and brown. This colour change is believed to be due to the conversion of chlorophyll to the compound pheophytin.

Conversion to pheophytin is favoured by acid pH but does not occur readily under alkaline conditions. For this reason peas, beans, spinach, and other green vegetables which tend to lose their bright green colours on heating can be largely protected against such colour changes by the addition of sodium bicarbonate or other alkali to the cooking or canning water.

However, this practice is not looked upon favourably nor used commercially because alkaline pH also has a softening effect on cellulose and vegetable texture and also destroys vitamin C and thiamin at cooking temperatures.

The Carotenoids. Pigments belonging to this group are fat-soluble and range in colour from yellow through orange to red. They often occur along with the chlorophylls in the chloroplasts, but also are present in other chromoplasts and may occur free in fat droplets. Important carotenoids include the orange carotenes of carrot, maize, apricot, peach, citrus fruits, and squash; the red lycopene of tomato, watermelon, and apricot; the yellow-orange xanthophyll of maize, peach, paprika and squash; and the yellow-orange crocetin of the spice saffron. These and other carotenoids seldom occur singly within plant cells.

A major importance of some of the carotenoids is their relationship to vitamin A. A molecule of orange beta-carotene is converted into two molecules of colourless vitamin A in the animal body. Other carotenoids like alpha-carotene, gamma-carotene, and cryptoxanthin also are precursors of vitamin A, but because of minor differences in chemical structure one molecule of each of these yields only one molecule of vitamin A.

In food processing the carotenoids are fairly resistant to heat, changes in pH, and water leaching since they are fat-soluble. However, they are very sensitive to oxidation, which results in both colour loss and destruction of vitamin A activity.

The Flavonoids. Pigments and colour precursors belonging to this class are water-soluble and commonly are present in the juices of fruit and vegetables. The flavonoids include the purple, blue, and red anthocyanins of grapes, berries, plump, eggplant, and cherry; the yellow anthoxanthins of light coloured fruit and vegetables such as apple, onion, potato, and cauliflower, and the colourless catechins and leucoanthocyanins which are food tannins and are found in apples, grapes, tea, and other plant tissues. These colourless tannin compounds are easily converted to brown pigments upon reaction with metal ions.

Properties of the anthocyanins include a shifting of colours with pH. Thus many of the anthocyanins which are violet or blue in alkaline media become red upon addition of acid.

Cooking of beets with vinegar tends to shift the colour from a purplish red to a brighter red, while alkaline water can influence the colour of red fruits and vegetables toward violet and gray-blue.

The anthocyanins also tend toward the violet and blue hues upon reaction with metal ions, which is one reason for lacquering the inside of metal cans when the true colour of anthocyanin-containing fruits and vegetables is to be preserved.

The water-soluble property of anthocyanins also results in easy leaching of these pigments from cut fruit and vegetables during processing and cooking.

The yellow anthoxanthins also are pH sensitive tending toward a deeper yellow in alkaline media. Thus potatoes or apples become somewhat yellow when cooked in water with a pH of 8 or higher, which is common in many areas. Acidification of the water to pH 6 or lower favours a whiter colour.

The colourless tannin compounds upon reaction with metal ions form a range of dark coloured complexes which may be red, brown, green, grey, or black. The various shades of these coloured complexes depend upon the particular tannin, the specific metal ion, pH, concentration of the complex, and other factors not yet fully understood.

Water-soluble tannins appear in the juices squeezed from grapes, apples, and other fruits as well as the brews from extraction of tea and coffee. The colour and clarity of tea are influenced by the hardness and pH of the brewing water. Alkaline waters that contain calcium and magnesium favour the formation of dark brown tannin complexes which precipitate when the tea is cooled.

If acid in the form of lemon juice is added to such tea its colour lightens and the precipitate tends to dissolve. Iron from equipment or from pitted tin cans has caused a number of unexpected colours to develop in products containing tannins, such as coffee, cocoa and foods flavoured with these.

The tannins are also important because they have an astringency which influences flavour and contributes body to such beverages as tea, wine, apple cider, etc.

Activities of Living Systems

Fruit and vegetables are in a live state after harvest. Continued respiration gives off carbon dioxide, moisture, and heat which influence storage, packaging, and refrigeration requirements. Continued transpiration adds to moisture evolved and further influences packaging requirements.

Further activities of fruit and vegetables, before and after harvest, include changes in carbohydrates, pectins, organic acids, and the effects these have on various quality attributes of the products.

As for changes in carbohydrates, few generalizations can be given with respect to starches and sugars. In some plant products sugars quickly decrease and starch increases in amount soon after harvest. This is the case for ripe sweet corn which can suffer flavour and texture quality losses in a very few hours after harvest.

Unripe fruit, in contrast, is frequently high in starch and low in sugars. Continued ripening after harvest generally results in a decrease in starch and a increase in sugars as in the case of apples and pears. However, this does not necessarily mean that the starch is the source of the newly formed sugars.

Further, the courses of change in starch and sugars are markedly influenced by postharvest storage temperatures. Thus potatoes stored below about 10 C° (50 F°) continue to build up high levels of sugars, while the same potatoes stored above 10 C° do not.

This property is used to help the dehydration process in potato storage. Here potatoes should have a low reducing sugar content so as to minimise Maillard browning reactions during drying and subsequent storage of the dried product. In this case potatoes are stored above 10°C prior to being further processed.

After harvest the pectin changes in fruit and vegetables are more predictable. Generally there is decrease in water-insoluble pectic substance and a corresponding increase in watersoluble pectin. This contributes to the gradual softening of fruits and vegetables during storage and ripening. Further breakdown of water-soluble pectin by pectin methyl esterase also occurs.

The organic acids of fruit generally decreases during storage and ripening. This occurs in apples and pears and is especially important in the case of oranges. Oranges have a long ripening period on the tree and time of picking is largely determined by degree of acidity and sugar content which have major effects upon juice quality.

Stability of Nutrients

One of the principal responsibilities of the food scientist and food technologist is to preserve food nutrients through all phases of food acquisition, processing, storage, and preparation. The key is in the specific sensitivities of the various nutrients, the principles of which are illustrated in Table 3.

Table 3. Specific sensitivity and stability of nutrients

Nutrient	*Neutral pH 7*	*Acid <pH 7*	*Alkaline >pH 7*	*Air or Oxygen*	*Light*	*Heat*	*Cooking Losses, Range*
Vitamins							
Vitamin A	S	U	S	U	U	U	0-40
Ascorbic acid (C)	U	S	U	U	U	U	0-100
Biotin	S	S	S	S	S	U	0-60
Carotenes	S	U	S	U	U	U	0-30
(pro A)							0-5
Choline	S	S	S	U	S	S	0-10
Cobalamin	S	S	S	U	U	S	
(B12)							0-40
Vitamin D	S		U	U	U	U	0-10
Essential	S	S	U	U	U	S	
Fatty acids							
Folic acid	U	U	S	U	U	U	0-100
Inositol	S	S	S	S	S	U	0-95
Vitamin K	S	U	U	S	U	S	0-5
Niacin (PP)	S	S	S	S	S	S	0-75
Pantbothenic acid	S	U	U	S	S	U	0-50
p-Amino	S	S	S	U	S	S	0-5
Benzoic acid							
Vitamin B6	S	S	S	S	U	U	0-40
Riboflavin (B2)	S	S	U	S	U	U	0-75
Tbiamin (B1)	U	S	U	U	S	U	0-80
Tocopherols	S	S	S	U	U	U	0-55

(U = Unstable; S = Stable)

This shows the stability of vitamins, essential amino acids, and minerals to acid, air, light, and heat, and gives an indication of possible cooking losses. Vitamin A is highly sensitive to acid, air, light and heat; vitamin C to alkalinity, air, light and heat; vitamin D to alkalinity, air, light and heat; thiamin to alkalinity, air, and heat in alkaline solutions; etc. Cooking losses of some essential nutrients may be in excess of 75%. In modern food processing operations, however, losses are seldom in excess of 25%.

The ultimate nutritive value of a food results from the sum total of losses incurred throughout its history - from farmer to consumer. Nutrient value begins with genetics of the plant and animal. The farmland fertilization program affects tissue composition of plants, and animals consuming these plants. The weather and degree of maturity at harvest affect tissue composition.

Storage conditions before processing affect vitamins and other nutrients. Washing, trimming, and heat treatments affect nutrient content. Canning, evaporating, drying, and freezing alter nutritional values, and the choices of times and temperatures in these operations frequently must be balanced between good bacterial destruction and minimum nutrient destruction.

Packaging and subsequent storage affect nutrients. One of the most important factors is the final preparation of the food in the home and the restaurant - the steam table can destroy much of what has been preserved through all prior manipulations.

Structural Features

The structural unit of the edible portion of most fruits and vegetables is the parenchyma cell. While parenchyma cells of different fruit and vegetables differ somewhat in gross size and appearance, all have essentially the same fundamental structure.

Parenchyma cells of plants differ from animal cells in that the actively metabolising protoplast portion of plant cells represents only a small fraction, of the order of five per cent, of the total cell volume. This protoplast is film-like and is pressed against the cell wall by the large water-filled central vacuole.

The protoplast has inner and outer semi-permeable membrane layers; the cytoplasm and its nucleus are held between them. The cytoplasm contains various inclusions, among them starch granules and plastics such as the chloroplasts and other pigment-containing chromoplasts. The cell wall, cellulose in nature, contributes rigidity to the parenchyma cell and limits the outer protoplasmic membrane. It is also the structure against which other parenchyma cells are cemented to form extensive three-dimensional tissue masses.

The layer between cell walls of adjacent parenchyma cells is referred to as the middle lamella, and is composed largely of pectic and polysaccharide cement-like materials. Air spaces also exist, especially at the angles formed where several cells come together.

The relationships between these structures and their chemical compositions are further outlined below. The parenchyma cells will vary in size among plants but are quite large when compared to bacterial or yeast cells. The larger parenchyma cells may have volumes many thousand times greater than a typical bacterial cell.

There are additional types of cells other than parenchyma cells that make up the familiar structures of fruit and vegetables. These include various types of conducting cells which are tube-like and distribute water and salts throughout the plant.

Such cells produce fibrous structures toughened by the presence of cellulose and the woodlike substance lignin. Cellulose, lignin, and pectic substances also occur in specialised supporting cells which increase in importance as plants become older.

An important structural feature of all plants, including fruit and vegetables is protective tissue. This can take many forms but usually is made up of specialised parenchyma cells that are pressed compactly together to form a skin, peel or rind.

Surface cells of these protective structures on leaves, stems or fruit secrete waxy cutin and form a water impermeable cuticle. These surface tissues, especially on leaves and young stems will also contain numerous valve-like cellular structures, the stomata, through which moisture and gases can pass.

Deterioration Factors and Their Control

A summary of overall deterioration reactions in fruits and vegetables is presented below.

Enzymic Changes

Enzymes which are endogenous to plant tissues can have undesirable or desirable consequences. Examples involving endogenous enzymes include a) the post-harvest senescence and spoilage of fruit and vegetables; b) oxidation of phenolic substances in plant tissues by phenolase (leading to browning); c) sugar - starch conversion in plant tissues by amylases; d) post-harvest demethylation of pectic substances in plant tissues (leading to softening of plant tissues during ripening, and firming of plant tissues during processing).

The major factors useful in controlling enzyme activity are: temperature, water activity, pH, chemicals which can inhibit enzyme action, alteration of substrates, alteration of products and pre-processing control.

Chemical Changes

Sensory quality

The two major chemical changes which occur during the processing and storage of foods and lead to a deterioration in sensory quality are lipid oxidation and non-enzymatic browning. Chemical reactions are also responsible for changes in the colour and flavour of foods during processing and storage.

Lipid oxidation rate and course of reaction is influenced by light, local oxygen concentration, high temperature, the presence of catalysts (generally transition metals such as iron and copper) and water activity. Control of these factors can significantly reduce the extent of lipid oxidation in foods.

Non-enzymic browning is one of the major causes of deterioration which occurs during storage of dried and concentrated foods. The non-enzymic browning, or Maillard reaction, can be divided into three stages: a) early Maillard reactions which are chemically well-defined steps without browning; b) advanced Maillard reactions which lead to the formation of volatile or soluble substances; and c) final Maillard reactions leading to insoluble brown polymers.

Colour changes

Chlorophylls. Almost any type of food processing or storage causes some deterioration of the chlorophyll pigments. Phenophytinisation (with consequent formation of a dull olivebrown phenophytin) is the major change; this reaction is accelerated by heat and is acid catalysed.

Other reactions are also possible. For example, dehydrated products such as green peas and beans packed in clear glass containers undergo photo-oxidation and loss of desirable colour.

Anthocyanins. These are a group of more than 150 reddish water-soluble pigments that are very widespread in the plant kingdom. The rate of anthocyanin destruction is pH dependent, being greater at higher pH values. Of interest from a packaging point of view is the ability of some anthocyanins to form complexes with metals such as Al, Fe, Cu and Sn.

These complexes generally result in a change in the colour of the pigment (for example, red sour cherries react with tin to form a purple complex) and are therefore undesirable. Since metal packaging materials such as cans could be sources of these metals, they are usually coated with special organic linings to avoid these undesirable reactions.

Carotenoids. The carotenoids are a group of mainly lipid soluble compounds responsible for many of the yellow and red colours of plant and animal products. The main cause of carotenoid degradation in foods is oxidation. The mechanism of oxidation in processed foods is complex and depends on many factors. The pigments may auto-oxidise by reaction with atmospheric oxygen at rates dependent on light, heat and the presence of pro- and antioxidants.

Flavour changes

In fruit and vegetables, enzymically generated compounds derived from long-chain fatty acids play an extremely important role in the formation of characteristic flavours. In addition, these types of reactions can lead to significant off-flavours. Enzyme-induced oxidative breakdown of unsaturated fatty acids occurs extensively in plant tissues and this yield characteristic aromas associated with some ripening fruits and disrupted tissues.

The permeability of packaging materials is of importance in retaining desirable volatile components within packages, or in permitting undesirable components to permeate through the package from the ambient atmosphere.

Nutritional quality

The four major factors which affect nutrient degradation and can be controlled to varying extents by packaging are light, oxygen concentration, temperature and water activity. However, because of the diverse nature of the various nutrients as well as the chemical heterogeneity within each class of compounds and the complex interactions of the above variables, generalizations about nutrient degradation in foods will inevitably be broad ones.

Vitamins. Ascorbic acid is the most sensitive vitamin in foods, its stability varying markedly as a function of environmental conditions such as pH and the concentration of trace metal ions and oxygen. The nature of the packaging material can significantly affect the stability of ascorbic acid in foods. The effectiveness of the material as a barrier to moisture and oxygen as well as the chemical nature of the surface exposed to the food are important factors.

For example, problems of ascorbic acid instability in aseptically packaged fruit juices have been encountered because of oxygen permeability of the package and the oxygen dependence of the ascorbic acid degradation reaction.

Also, because of the preferential oxidation of metallic tin, citrus juices packaged in cans with a tin contact surface exhibit greater stability of ascorbic acid than those in enamelled cans or glass containers. The aerobic and anaerobic degradation reactions of ascorbic acid in reduced-moisture foods have been shown to be highly sensitive to water activity, the reaction rate increasing in an exponential fashion over the water activity range of 0.1-0.8.

Physical Changes

One major undesirable physical change in food powders is the absorption of moisture as a consequence of an inadequate barrier provided by the package; this results in caking. It can occur either as a result of a poor selection of packaging material in the first place, or failure of the package integrity during storage. In general, moisture absorption is associated with increased cohesiveness.

Anti-caking agents are very fine powders of an inert chemical substance that are added to powders with much larger particle size in order to inhibit caking and improve flowability. Studies in onion powders showed that at ambient temperature, caking does not occur at water activities of less than about 0.4.

At higher activities, however, ($aw > 0.45$) the observed time to caking is inversely proportional to water activity, and at these levels anti-caking agents are completely ineffective. It appears that while they reduce inter-particle attraction and interfere with the continuity of liquid bridges, they are unable to cover moisture sorption sites.

Biological Changes

Microbiological

Micro-organisms can make both desirable and undesirable changes to the quality of foods depending on whether or not they are introduced as an essential part of the food preservation process or arise unintentionally and subsequently grow to produce food spoilage.

The two major groups of micro-organisms found in foods are bacteria and fungi, the latter consisting of yeasts and moulds. Bacteria are generally the fastest growing, so that in conditions favourable to both, bacteria will usually outgrow fungi.

Foods are frequently classified on the basis of their stability as non-perishable, semi-perishable and perishable. For example, hermetically sealed and heat processed (e.g. canned) foods are generally regarded as non-perishable. However, they may become perishable under certain circumstances when an opportunity for recontamination is afforded following processing.

Such an opportunity may arise if the can seams are faulty, or if there is excessive corrosion resulting in internal gas formation and eventual bursting of the can. Spoilage may also take place when the canned food is stored at unusually high temperatures: thermophilic spore-forming bacteria may multiply, causing undesirable changes such as flat sour spoilage.

Low moisture content foods such as dried fruit and vegetables are classified as semi-perishable. Frozen foods, though basically perishable, may be classified as semi-perishable provided that they are properly stored at freezer temperatures.

The majority of foods (e.g. meat and fish, milk, eggs and most fresh fruits and vegetables) are classified as perishable unless they have been processed in some way. Often, the only form of processing which such foods receive is to be packaged and kept under controlled temperature conditions.

The species of micro-organisms which cause the spoilage of particular foods are influenced by two factors: a) the nature of the foods and b) their surroundings. These factors are referred to as intrinsic and extrinsic parameters.

The intrinsic parameters are an inherent part of the food: pH, a_w, nutrient content, antimicrobial constituents and biological structures. The extrinsic parameters of foods are those properties of the storage environment that affect both the foods and their microorganisms. The growth rate of the micro-organisms responsible for spoilage primarily depends on these extrinsic parameters: temperature, relative humidity and gas compositions of the surrounding atmosphere.

The protection of packaged food from contamination or attack by micro-organisms depends on the mechanical integrity of the package, and on the resistance of the package to penetration by micro-organisms.

Metal cans which are retorted after filling can leak during cooling, admitting any microorganisms which may be present in the cooling water, even when the double seam is of a high quality. This fact is widely known in the canning industry and is the reason for the mandatory chlorination of cannery cooling water. Extensive studies on a variety of plastic films and metal foils have shown that microorganisms (including mounds, yeasts and bacteria) cannot penetrate these materials in the absence of pinholes.

In practice, however, thin sheets of packaging materials such as aluminium and plastic do contain pinholes. There are several safeguards against the passage of micro-organisms through pinholes in films:

— because of surface tension effects, micro-organisms cannot pass through very small pinholes unless the micro-organisms are suspended in solutions containing wetting agents and the pressure outside the package is greater than that within;

— materials of packaging are generally used in thicknesses such that pinholes are very infrequent and small;

— for applications in which package integrity is essential (such as sterilisation of food in pouches), adequate test methods are available to assure freedom from bacterial recontamination.

Macrobiological

Insect Pests

Warm humid environments promote insect growth, although most insects will not breed if the temperature exceeds about 35 C° or falls below 10 C°. Also many insects cannot reproduce satisfactorily unless the moisture content of their food is greater than about 11%.

The main categories of foods subject to pest attack are cereal grains and products derived from cereal grains, other seeds used as food (especially legumes), dairy products such as cheese and milk powders, dried fruits, dried and smoked meats and nuts.

As well as their possible health significance, the presence of insects and insect excrete in packaged foods may render products unsaleable, causing considerable economic loss, as well as reduction in nutritional quality, production of off-flavours and acceleration of decay processes due to creation of higher temperatures and moisture levels.

Early stages of infestation are often difficult to detect; however, infestation can generally be spotted not only by the presence of the insects themselves but also by the products of their activities such as webbing, clumped-together food particles and holes in packaging materials.

Unless plastic films are laminated with foil or paper, insects are able to penetrate most of them quite easily, the rate of penetration usually being directly related to film thickness. In general, thicker films are more resistant than thinner films, and oriented films tend to be more effective than cast films. The looseness of the film has also been reported to be an important factor, loose films being more easily penetrated than tightly fitted films.

Generally, the penetration varies depending on the basic resin from which the film is made, on the combination of materials, on the package structure, and of the species and stage of insects involved. The relative resistance to insect penetration of some flexible packaging materials is as follows:

— *excellent resistance*: polycarbonate; poly-ethylene-terephthalate;

— *good resistance:* cellulose acetate; polyamide; polyethylene (0.254 mm); polypropylene (biaxially oriented); poly-vinyl-chloride (unplasticised);

— *fair resistance*: acrylonitrile; poly-tetra-fluoro-ethylene; polyethylene (0.123 mm);

— *poor resistance*: regenerated cellulose; corrugated paper board; kraft paper; polyethylene (0.0254 - 0.100 mm); paper/foil/polyethylene laminate pouch; poly-vinylchloride (plasticised).

Some simple methods for obtaining insect resistance of packaging materials are as following:

— select a film and a film thickness that are inherently resistant to insect penetration;
— use shrink film over-wraps to provide an additional barrier;
— seal carton flaps completely.

Rodents

Rats and mice carry disease-producing organisms on their feet and/or in their intestinal tracts and are known to harbour salmonella of serotypes frequently associated with food-borne infections in humans. In addition to the public health consequences of rodent populations in close proximity to humans, these animals also compete intensively with humans for food.

Rats and mice gnaw to reach sources of food and drink and to keep their teeth short. Their incisor teeth are so strong that rats have been known to gnaw through lead pipes and unhardened concrete, as well as sacks, wood and flexible packaging materials.

Proper sanitation in food processing and storage areas is the most effective weapon in the fight against rodents, since all packaging materials apart from metal and glass containers can be attacked by rats and mice.

Methods of Reducing Deterioration

A knowledge of deterioration factors and the way they act, including the rates of deterioration to a specific category of food, means that it is possible to list the ways of lowering or stopping the action and obtaining fruit and vegetable preservation.

In order to maintain their nutritional value and organoleptic properties and because of technical-economical considerations, not all the identified means against deterioration actually have practical applications for fruit and vegetable preservation.

Technical Methods of Reducing Food Deterioration

These technical means can be summarised as follows:

Physical	Heating
	Cooling
	Lowering of water content Drying/dehydration. Concentration

	Sterilising filtration
	Irradiation
	Other physical means (pressure, vacuum, inert gases)
Chemical	Salting
	Smoking
	Sugar addition
	Artificial acidification
	Ethyl alcohol addition
	Antiseptic substance action
Biochemical	Lactic fermentation (natural acidification)
	Alcoholic Fermentation

This classification of methods of reducing deterioration presents some difficulties because their preservation effects are physical, physico-chemical, chemical and biochemical complex phenomena which rarely act in isolation. Normally they take place together or one after the other.

From the whole list of possible methods of reducing deterioration, over the years, some procedures for fruit and vegetable preservation have found practical application.

Table 4. Procedures for fruit and vegetable preservation

Procedures	*Practical applications*
Fresh storage	Fruits, vegetables
Cold storage	Fruits, vegetables
Freezing	Fruits, vegetables
Drying/dehydration	Fruits, vegetables
Concentration	Fruit and vegetable juices
Chemical preservation	Fruit semi-processed
Preservation with sugar	Fruit products/preserves
Pasteurization	Fruit and vegetable juices
Sterilisation	Fruits, vegetables
Sterilising filtration	Fruit juices
Irradiation	Fruits, vegetables

These preservation procedures have two main characteristics as far as being applied to all food products is concerned:

— some of them are applied only to one or some categories of foods; others can be used across the board and thus a wider application (cold storage, freezing, drying/dehydration, sterilisation, etc.);

— some guarantee food preservation on their own while others require combination with other procedures, either as principal or as auxiliary processes in order to assure preservation (for example smoking has to be preceded by salting).

Combined Preservation Procedures

In practice preservation procedures aim at avoiding microbiological and biochemical deterioration which are the principal forms of deterioration. Even with all recent progress achieved in this field, no single one of these technological procedures applied alone can be considered wholly satisfactory from a microbiological, physico-chemical and organoleptic point of view, even if to a great extent the food value is assured.

Thus, heat sterilisation cannot be applied in order to destroy all micro-organisms present in foods without inducing non desirable modifications. Preservation by dehydration/drying assures microbiological stability but has the drawback of undesirable modifications that appear during storage: vitamin losses, oxidation phenomena, etc.

Starting with these considerations, the actual tendency in food preservation is to study the application of combined preservation procedures, aiming at the realisation of maximum efficiency from a microbiological and biological point of view, with reduction to a minimum of organoleptical degradation and decrease in food value.

The principles of combined preservation procedures are:

— avoid or reduce secondary (undesirable) effects in efficient procedures for microbiological preservation;

— avoid qualitative degradation appearing during storage of products preserved by efficient procedures from a microbiological point of view;

— increase microbiological efficiency of preservation procedures by supplementary means;

— combine preservation procedures in order to obtain maximum efficiency from a microbiological point of view, by specific action on various types of micro-organisms present;

— establish combined factors that act simultaneously on bacterial cells.

Research and applications in this direction were followed by microbiological and biochemical way, obtaining a serial of combination of preservation procedures with the possibility of application in industrial practice.

Fresh fruit and vegetable storage can be combined with:

— storage in controlled atmosphere where carbon dioxide and oxygen levels are monitored, increasing concentration of CO_2 and lowering that of oxygen according to fruit species. Excellent results were obtained for pomace fruit; in particular the storage period for apples has been extended. Application of this combined procedure requires airtight storage rooms.

— storage in an environment containing ethylene oxide; this accelerates ripening in some fruit: tomatoes, bananas, mangoes, etc.

Cold storage can be combined with storage in an environment with added of carbon dioxide, sulphur dioxide, etc. according to the nature of product to be preserved.

Preservation by drying/dehydration can be combined with:

— *freezing*: fresh fruit and vegetables are dehydrated up to the point where their weight is reduced by 50% and then they are preserved by freezing.

This procedure (freeze-drying) combines the advantages of drying (reduction of volume and weight) with those of freezing.

A significant advantage of this process is the short drying time in so far as it is not necessary to go beyond the inflexion point of the drying curve. The finished products after defreezing and rehydration/reconstitution are of a better quality compared with products obtained by dehydration alone.

— cold storage of dried/dehydrated vegetables in order to maintain vitamin C; storage temperature can be varied with storage time and can be at -8° C for a storage time of more than one year, with a relative humidity of 70-75 %.

— packaging under vacuum or in inert gases in order to avoid action of atmospheric oxygen, mainly for products containing beta-carotene.

— chemical preservation: a process used intensively for prunes and which has commercial applications is to rehydrate the dried product up to 35 % using a bath containing hot 2 % potassium sorbate solution. Another possible application of this combined procedure is the initial dehydration up to 35% moisture followed by immersion in same bath as explained above; this has the advantage of reducing drying time and producing minimum qualitative degradation. Both applications suppress the dehydrated products reconstitution (rehydration) step before consumption.

— packaging in the presence of desiccants (calcium oxide, anhydrous calcium chloride, etc.) in order to reduce water vapour content in the package, especially for powdered products.

Preservation by concentration, carried out by evaporation, is combined with cold storage during warm season for tomato paste (when water content cannot be reduced under the limit needed to inhibit moulds and yeasts, e.g. $a_w = 0.70...0.75$).

Chemical preservation is combined with:

— acidification of food medium (lowering pH);

— using combined chemical preservatives.

Preservation by lactic fermentation (natural acidification) can be combined with cold storage for pickles in order to prolong storage time or shelf-life.

Preservation with sugar is combined with pasteurization for some preserves having a sugar content below 65%.

References

Colin, D. 1992. Recent trends in fruit and vegetable processing. In *Food Science and Technology Today*, 7, (2)), pp. 111 - 116.

Desrosier, N.W. and Desrosier, J.N. 1977. *Technology of Food Preservation*, 4th Edition, AVI Publishing Co. Westport, Conn.

FAO. 1986. *Prevention of post-harvest food losses.* Rome: FAO.

Harris, R.Sa And Karmas, E. 1975. *Nutritional Evaluation of Food Processing*. 2nd Edition.AVI Publishing Co.,Westport,Conn.

2

Infrastructure of Processing Plant

When considering the setting up of a fruit and vegetable processing plant, whether it be a cottage industry or a small industrial scale system, the first point to bear in mind is the infrastructure required to properly lodge all of the necessary equipment.

Some time must thus be devoted to coordinate two aspects that are vital to the development of a project of this nature, namely costs and the quality of the infrastructure needed to achieve the established goals.

It must always be borne in mind that since the food to be processed is intended for human consumption, the infrastructure must meet several requirements.

The infrastructure comprises different aspects of a project's implementation. Issues like physical layout, basic services or installations and equipment must thus be taken into account.

Physical Layout

The physical layout of a plan of this nature may be very simple, as it refers to a basic production system, involving small volumes and simple products, from a technological point of view.

Nevertheless, in the case of a cottage industry and a small industrial scale system alike, simplicity must never neglect the basic principles governing industrial health and hygiene, which must characterize a food production system.

Production Sites

Several different processes take place on the site where the production activity is performed, from the reception and conservation of raw materials, to the storage of finished products.

One aspect that must be borne in mind relates to construction details, which determine a plant's capacity to meet two objectives: to adjust to the production of foods and to ensure a sufficiently long shelf life. However, when considering home or small-scale industrial processing facilities, the cost of construction is an important factor which must be taken into account.

The building materials must be as light as possible, easy to readapt and install, considering that often the system users develop the plan themselves, by means of self-construction methods.

The buildings materials must be easy to readapt because these home-made systems are rather dynamic, that is, they require frequent changes or must adjust to different processes, so that the space that they occupy may be exploited all year round. On the other hand, these systems must be considered "expandable" to accommodate possible evolutions in time.

In addition to the previously mentioned characteristics, the materials must be easy to wash and disinfect, especially those in the clean areas of the processing rooms. Complex type of construction, resulting in the creation of places that are not easily accessible for cleaning must be avoided, for they may turn into bird nests, and contamination foci for rodents, insects, and of course, micro-organisms.

Requirements pertaining to the materials and construction characteristics of the sites do not vary greatly for home processing or small-scale industrial plants. The basic difference lies in the equipment and the way it is set up in the processing lines. The home-processing system is temporary and versatile, and there are no special areas devoted to a single process. In general, all of the premises serve several purposes, according to the type of process and raw material being used.

The small-scale industrial system, on the other hand, is more complex in its organization, and therefore specific activities are carried out in determined areas. Nevertheless, the general requirements for both systems are similar, the difference being in the way such requirements are met.

Some of the aspects that may be considered important in relation to the architectural and construction elements are listed below:

The ceiling and walls of the processing room must be of washable and easily dried materials; they must be neither absorbent nor porous.

— The lighting should be natural, as far as possible. However, if artificial lights must be used, they should not hinder activities in any way. Artificial lighting must be protected, to prevent fragments of glass from falling into the product as it is being processed, in case of accidents.

— Ideally, the working environment should always be appropriately ventilated, to facilitate the workers' performance. Poor ventilation in highly enclosed and densely populated premises may generate defects. It is also important to provide for the elimination of heavily contaminating odours, even if they are not necessarily toxic.

On the other hand, excess ventilation, especially in places characterized by great aerial contamination external to the processing site, dust and insects essentially, may prove to be counterproductive. Appropriate ventilation must therefore be based on an efficient system controlling the access of foreign material from the external environment.

— The floors must be of a solid material, never earth or plant covering. Like the walls and ceiling of the processing room, the floor must be washable, to ensure compliance with the premises' hygienic and health standards. The floor must also be sloped to allow appropriate drainage, avoiding at all costs the formation of pools in the processing area. At the same time, care must be taken to prevent the floor from being slippery.

These are some examples of the features that must characterize a fruit and vegetable processing site to guarantee a quality product suitable for human consumption.

Basic Installations or Services

Three basic services are required for the operation of a system as the one in question: electrical power, drinking water and the disposal of waste waters.

Occasionally, small-scale industrial plants are equipped with a steam production system, which however is more seldom found in home-processing plants.

Even when a home-processing plant can operate without electrical power, it is better for this service to be available, essentially to facilitate the processes by means of small devices that were developed and that improve workers' performance, thus guaranteeing a greater uniformity of products.

Electrical power is also absolutely necessary if one is to rely on an appropriate lighting system, so that work shifts may be prolonged, especially when there is a surplus production of raw materials.

In small-scale industrial production systems, electrical energy is indispensable, due to the greater degree of mechanization of the processes involved. All lights must be installed on the ceiling at a safe distance to prevent them from getting wet and getting in the way of workers in the processing room.

As to water supply, the problem is slightly more critical. Sufficient drinking water must be available to ensure the development of a hygienic process, managed by clean people and with appropriately disinfected equipment. Also, many processes require water, as a result of which water of an appropriate quality must be available.

Since water does not come in abundant quantities, its use must therefore be regulated by strict savings principles, especially in small or home-processing installations that normally are not equipped with sophisticated water harnessing devices. Water must be protected from possible sources of contamination and must be supplied on a continuous basis at all times. The consumption of water will depend upon the process in question and the design of the production systems.

The supply of water must be ensured on a permanent basis, as a result of which the plant will need to be equipped with an elevated storage tank to avoid being dependent on the supply of electricity. A reserve must be created, so that water is available even when there is no electrical power. Tank storage will also allow for treatment through the addition of disinfectants.

In general, it is advised that chlorine be added to the water supplying the entire plant, so as to provide for permanent disinfection. To this end, a dose of 2 ppm of residual free chlorine is suggested. It should also be borne in mind that the tank must be covered and not exposed to sunlight, to prevent the chlorine from decomposing. As a term of reference, 100 ml of a sodium hypochlorite solution for every 2000 liters of water may be used, assuming that the hypochlorite solution contains about 50 mgr of active chlorine per litre of solution. This will prevent the water from having any chlorine-like taste.

Basic Facilities

A fruit and vegetable processing plant must be set up in such a way as to rely on a number of basic facilities, which are generally similar in home-processing and small-scale industrial systems.

Reception of Raw Material

The plant must be equipped with a special area for the reception of raw materials, that is, a site where the raw material received in appropriate conditions may be stored until it is used in the process. This site, which may simply be a shed or a more appropriately designed room, must meet certain special standards in terms of temperature, humidity cleanliness, and exposure to sunlight. It is important to consider that the quality of most raw materials covered in this manual rapidly deteriorates. That is, even though many species do preserve their integrity, their inner quality is subjected to variations if storage conditions are less than adequate.

It is for this reason that the temperature must be as low as possible; it must be cool. The raw material must not be directly exposed to sunlight. Since storage temperature is a very important factor, if a refrigeration system is not available, the material must be collected in the cool hours of the day.

If the storage site is cool, it is important for the humidity to be relatively high to prevent the material from dehydrating and losing its quality. This problem does not apply to areas with a high relative humidity, in which case the only requirement is to find a cool site.

It is important to underscore that the raw material storage area must not be used for the storage of other products that may be contaminating, such as pesticides, paint, or cleaning utensils, all of which must be kept in specially designated areas.

It must never be forgotten that the quality of the product will reflect the quality of the raw material from which it was made; it is therefore important to take this aspect into due account.

This storage site must be provided with basic equipment for the reception of the material. The scale and other instruments for primary quality control must be kept in a safe place, where they will not be damaged. An appropriate place must have an average temperature no higher than 30°C and a humidity no greater than 70%. The tools must be kept in their respective cases at all times, clean and dry.

Processing room

The processing room is the main facility in a plant of this type. It is here that the different materials used in the processing of the raw material are stored. On such premises, a continuous production line may be set up, or simply an ensemble of small machines allowing the products to be processed by hand and on a discontinuing basis. Ideally, this room should be big enough to lodge all of the necessary equipment on a continuous line, even in barely automated facilities. Even in the case of work benches where the work is performed by hand, the process must be carried out on the basis of a continuous line, to step up efficiency.

The processing room should ideally be divided into areas where different functions are performed. This may be achieved by separating such areas physically. Generally, there is a "dirty" area, that is, an area where the raw material is washed and peeled, and where operations like pitting, coring, and the removal of inedible parts are performed. This "dirty" area must not extend to the section of the plant or of the processing room where the cleanest operations are carried out, like pulp extraction, grinding, cutting and the filling of containers.

One way of achieving this separation is through the use of light partitions, painted wood panels used to simply separate one area from the other. Much care should be taken to avoid contamination by run-off waters. The recontamination of materials that have already been washed and disinfected is a common problem in home or small-scale industrial processing plants.

Quality control

Ideally, quality control operations should be performed in small quarters, which may also be separated from other areas by wood panels, where the basic tests required to establish the quality of a given raw material or a given process may be performed. This area should preferably be equipped with a small sink, running water and a counter where tests may be carried out.

It should be separated from the other quarters so that basic calculations may be carried out in a quiet environment.

Storeroom for finished products

The storeroom is fundamental in a plant of this type. It is often necessary for the product to remain under observation before being consumed.

Sometimes, the products must settle for a while to achieve a certain level of homogeneity, whereas in other cases the material must await labelling. Finally, in addition to being able to rely on a room where the material may be safely stored, it is also necessary to have access to a site where the process may be completed. Such a place must be clean, the temperature and humidity levels must be appropriate (less than 25 C° and 60% of relative humidity), and it must be protected from foreign matter, and naturally, from thieves. It should be easily accessible, so that tests may be performed during product storage, and any problems may be detected on the spot.

Other facilities

Some equipment, due to its nature, cannot be installed in the main facility of a processing plant. The boiler is an example. If the plant is equipped with a small steam generator, it should be located outside the processing room, to avoid contamination problems, and at the same time ensure personnel safety.

A drier is another special system, which should be installed in a rather dry place and not in the processing room, as this is an especially humid area in the plant.

Dehydrated products should normally be very low in moisture, a condition that can only be fulfilled if dehydration is carried out in an especially dry place, even if an artificial drier is employed. Otherwise, the energy consumption cost will be very high, since a great amount of heat will be required to dry the air.

Sanitary facilities

Sanitary facilities are believed to deserve special mention, due to the significant role that they play in preserving health and sanitary standards in a plant of this type.

The conditions in which the sanitary facilities operate, the type of evacuation system serving the plant, the location of the facilities and the sanitation plan are crucial to the quality of the process.

One basic condition is for the facilities to be erected in a separate location from the area where the raw material is received and processed, to prevent possible flooding. The facilities must be periodically disinfected, and the firm's supervisors must exercise very strict control in this regard.

It should be borne in mind that even though the current cholera outbreak in Latin America is viewed as an isolated case, health care should not be a priority in times like these alone.

Indeed, there is always some micro-organism around that may be detrimental to the health of whoever consumes the product.

Sanitary facilities must never be short of water. Its supply must be guaranteed, since the cleanliness of the toilets will determine the cleanliness of the workers, and the products' sanitary qualities will ultimately depend on the cleanliness of the workers.

Equipment

Figures 1 to 6 illustrates different implements and machines comprising the basic equipment required for the home processing of fruit and vegetables. Figure 2 shows a steam-powered heating system, figure 3 shows a press for the extraction of juice, and figure 4 shows a pulp removing machine.

The most common processes that apply to fruit are drying, preservation, pulp concentration, the manufacturing of juice, nectars and sweets, and concentrated pulp processing.

Home Processing Equipment

Figure 1 to 4 illustrate milling systems. In the first case, a pulp extractor used for fruit as well as tomatoes and vegetables is shown. It is provided with a sieve to separate the seeds and skin from the juice, which is the basic raw material to be used in the process.

Figure 1. Electrically powered pulp extractor

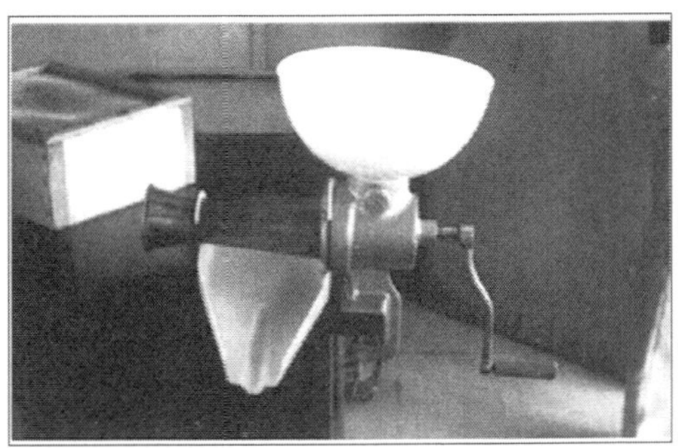

Figure 2. Manual pulp extractor

Figure 3. Hand mill

Figure 4. Commonhand sieve

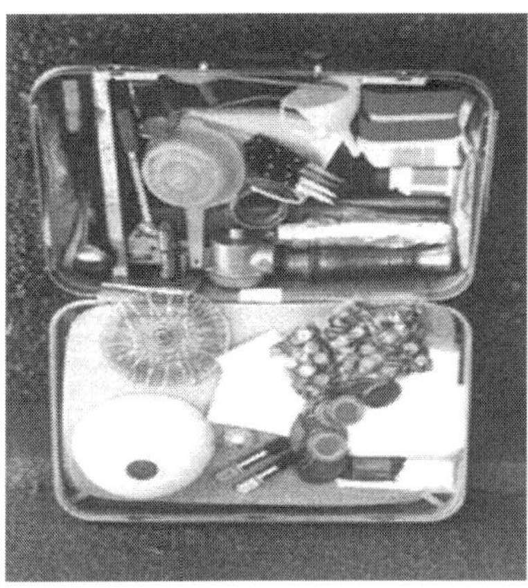

Figure 5. Kit containing the equipment and demonstration materials

Figure 6. Movable solar drier

Figure 4 shows a common hand powder separating sieve.

Figure 5 shows a multiple use kit containing a series of materials for fruit processing. This kit is used for training courses, but it contains all of the elements which, on a larger scale may constitute the basis for the home processing of various fruits and vegetables.

Specifications for the building or adaptation of industrial premises

— A processing area (5(10) x 10 m approx.) possibly equipped with a ceiling fan, a mosquito-net and a room in which to store packaging material, additives and finished products (4 x 4 m). Ample natural and artificial lighting.

— Sanitary facilities outside the processing area.

— Electrical power supply, and to the extent possible, sockets on each wall of the processing area, high up above the ground and away from the humid floor.

— Double dishwasher, preferably enamelled or of stainless steel, with running drinking water.

— Two double gas stoves, with their respective cylinders and regulators. As an alternative, electrical, paraffin or firewood-generated heat may be used.

— Drinking water (in the processing area and surroundings).

— Two enamelled or painted wood tables (180 x 120 x 80 cm approximately), with a galvanized steel or ideally a stainless steel covering.

Materials

— Bottles with crown cork. As an alternative, use between 500 and 1,000 disposable or returnable beer bottles (of approximately 200-280 ml).

— Between 2,000 and 5,000 metal crown cork.

— 500 glass jars (of 450 gr approximately) with screw-on or twist-off lids.

— 200 glass jars (of 900 gr approximately) with screw-on or twist-off lids.

— Screw-on or twist-off lids for jars of different sizes.

— Adhesive labels for bottles and jars.

— Citric acid, 500 gr, or lemon juice, 3 litres.

— Pectin powder for foods, 2 kg.

— Refined sugar, the amount of which will depend on the volume of the product to be obtained.

— 10 empty sacks of flour (1m x 0.5 m approximately).

— 1 kg of sodium benzoate for foods, optional.

— 1 kg of potassium sorbate for foods, optional.

— 1 kg of sodium metabisulfate, optional.

— Caustic soda.

Equipment

— Scale (from 50 to 100 kg).

— Scale (from 3 to 5 kg).
— Scale (from 100 to 500 gr).
— Hand refractometer (0 - 90° Brix)
— Refractometer (0 - 30° Brix)
— Stainless steel thermometer (0 - 150°C)
— 2 cast aluminium pots with lid (with a capacity of approximately 50 litres).
— 2 cast aluminium pots with lid (with a capacity of approximately 10 litres).
— 2 cast aluminium pots with lid (with a capacity of approximately 5 litres).
— 10 wooden chopping boards (40 x 30 cm).
— 5 stainless steel knives with a thick blade (15-20 cm x 2 cm).
— 5 stainless steel knives with a thick blade (10 cm x 1 cm).
— 5 colanders (25-20 cm diameter) with aluminium mesh.
— 5 plastic trays (40 x 60 x 5 cm).
— 10 plastic buckets (20 litres).
— 10 plastic buckets (10 litres).
— 2 plastic or aluminium funnels (20 cm diameter).
— 2 plastic or aluminium funnels (15 cm diameter).
— 3 stainless steel spoons of different sizes.
— 3 large plastic spoons.
— 3 medium wooden spoons.
— 3 large wooden spoons.
— 2 manual pulp extractors/separators.
— 2 manual cappers for crown cork.
— 5 perforated plastic cases for fruit for 18-20 kg.

Equipment for a Small-scale Industrial Plant

When analyzing the equipment required by a small-scale industrial plant, it may be observed that there are no great differences in terms of basic principles.

The difference essentially lies in the size and application of electrical and mechanized equipment of a greater unit capacity, probably characterized by a greater resistance and durability, but based on the same technological principles.

In the specific case of a semi-industrial plant, the pots will be replaced by steam kettles, heating will be provided by a steam-boiler, and a small press as well as an autoclave will be available. A list of additional equipment that must be installed in a small-scale industrial plant is provided below.

— A small boiler producing 250 kilos of steam per hour.

— A vertical autoclave with a capacity for about 200-500 g jars.

— A pulper, which operates manually or electrically.

— A hand-operated hydraulic press.

— A pressure bottles closer.

— Two double-wall kettles.

Figure 5 illustrates the procedures involved in the preservation and processing of concentrates, in which the handling takes place on a larger scale as compared to home processing. It may also be observed that with the exception of the vacuum evaporator, the rest of the equipment is rather similar to that analyzed in the previous section, with a significant difference in size but governed by the same principles.

Since it is larger, automated to some extent and characterized by a greater use of electricity, a small-scale system requires installations in better conditions than a home processing system, although such requirements are significant in terms of space only.

Staff

In food manufacturing, the staff are the most important resource in the production process. This is equally true of home and small-scale industrial processing operations, in spite of the fact that they are essentially self-managed.

When speaking of home processing, it is immediately assumed that none of the staff are hired on the basis of regular employment conditions. The situation thus must be analyzed from a different perspective than that normally adopted for an activity of this kind.

The purpose of these considerations is to stress that even in home or small-scale industrial processing operations, or in the case of a small group of people, there is a value that must be duly assessed, namely the labour involved in the process. It should be borne in mind that even in the case of small businesses, there may be the traditional division between temporary and permanent staff. The permanent staff will undoubtedly comprise the entrepreneurs, those who are financially responsible and have set up the small or very small firm. The temporary staff will comprise workers hired on a provisional basis, as a result of a seasonal surplus of raw material, who will come to be part of a process that does not involve them financially or emotionally.

The permanent staff are responsible for the business, and have the greatest interest in the production activity. They are usually involved in the enterprise on both an emotional and financial level. In this type of firm, the permanent staff generally include the owners, those who conceived the idea of setting up the business, the developers, the sales people and the innovators. It is extremely important for this staff to receive training in the area of management as well as technology, although emphasis should be placed on management.

The training of temporary workers should especially focus on technology, it should be demonstrated to them how important it is to do things right from the beginning. It should be borne in mind that in a home or small-scale industrial processing system, the incidence of labour is of vital importance in the enterprise's financial position.

Industrial Sanitation and Health

General Sanitary Standards

Quality and health standards and regulations must be strictly applied, or the product will be exposed to contamination by bacteria, mould and yeasts, thus jeopardizing the expected development of an agroindustrial enterprise.

Such measures must be adopted as early as in the production phase, and must continue in the post-harvest, transportation, storage, preparation and processing phases.

In line with these principles, the following sanitary standards must be fulfilled and applied by workers on the production premises:

— Workers must wash their hands and clean their nails carefully before engaging in any process. They must keep their nails short, and if possible, use rubber gloves.

— To enter the working area, workers must wear a clean smock, a hair net to protect the food from possible contamination by hair, and a mask to avoid microbial contamination.

— The working utensils and equipment must be cleaned appropriately to remove any waste or residual organic material.

— The containers (glass jars and bottles) must be washed with hot water before being filled with food.

— The waste generated by the production process must be removed from the production area on a daily basis.

— Clean and dry the outside of the containers with the product before labelling and storing.

— The storage site of the finished product must be clean and free from all possible contamination (it must have been previously fumigated). It must also be cool and dry.

— Once the working cycle has been completed, the production area must be left perfectly clean. It will therefore have to be pre-rinsed with water at a temperature of 40°C (to remove about 90% of the dirt), washed with detergent, and finally rinsed with water at a temperature of 38-46°C.

— Both the premises and the equipment will have to be disinfected on a fortnightly basis. Caustic soda will be applied first (2%), and then nitric acid (1.5%) at a temperature of 75 0 C after which they will be rinsed with water.

Industrial Health Standards

Whereas hygiene is a principle that applies to people, industrial health applies to the equipment, facilities and premises utilized in the production process. It is extremely important to adopt measures to ensure that the facilities meet the industrial health standards which guarantee an efficient implementation of the process.

These standards apply to small, medium and large enterprises and cottage industries alike, and should also be applied at the home level.

— The buildings must be adjusted so that they can be easily cleaned. There should be no blind spaces inaccessible to the cleaning and disinfection system.

— The equipment must be designed in such a way that no empty spaces are left to facilitate the accumulation of material that may decompose and cause severe contamination problems.

— All surfaces exposed to food must be properly cleaned and disinfected, with a frequency that will depend upon the type of raw material and process being used. Fruit and vegetable residues are generally easy to clean.

— A disinfection process can never be performed on a dirty surface. In order for the disinfection process to be successful, the surface must have been cleaned beforehand.

— The products used both in the cleaning and disinfection processes must be included in the list of products authorized by local health authorities. Special care must be taken to avoid polluting the environment by using products with an uncertain degradability.

— No disinfection process by itself will ever be able to replace the need for daily compliance with general sanitary requirements.

Microbiological Tests

To emphasize and help the staff of a food processing firm understand the importance of industrial sanitation and health, a simple test may be performed. All one needs is the cooperation of a centre equipped for the preparation of Petri dishes with a general-purpose microbiological culture medium like agar-starch-dextrose, and their subsequent incubation.

A certain number of Petri dishes are prepared and sterilized. If there is no university centre or a similar institution in the area, a local hospital may be asked to collaborate in the performance of this task. The dishes must be "seeded" with different elements that may constitute the source of possible microbiological contamination.

These may include the workers' hands, nails, hair or shoe soles (a smear is made in this case), the counters in the processing area (SMEAR), skin from the face of staff members (especially in the area near the mouth and nose), the air in the environment, the water utilized in the process and other elements that one might wish to analyze.

A smear is made by rubbing a sterilized cotton swab on the area to be analyzed and then on the agar surface. The cotton ball is normally placed on a wooden stick 10 cm long and 2-3 mm thick, so that it forms a small brush.

After being inoculated, the dishes are properly marked, sealed, and are left to incubate at a temperature between 25 and 30°C. Qualitative findings will be observable after 4-6 days, and will be sufficient for a general analysis of the working environment.

It is interesting to observe that the particles collected from the workers' fingertips in the illustrated case did not elicit the development of microorganisms. The same may be said for the particles collected from the counters and the workers' hair, which is indicative of excellent sanitary and health conditions. On the other hand, it is important to observe that the water used in the tests. This is an important finding, as the water used was drinking water drawn from the general water supply system serving the institution involved, which was fed by a supposedly sterile deep well.

It is also interesting to note the contamination developed by skin particles, which in this case were collected from the forehead and the nose region of a production line woman worker.

The air from the environment was also found to be highly contaminated, probably by dust, unlike the sample containing particles from the cough of a worker with incipient influenza. Nail particles showed a small degree of micro-organism growth.

A similar test may be carried out periodically to monitor the general level of the working environment and to raise the awareness of the staff as to the need for personal as well as plant hygiene.

Industrial Agricultural Production Unit

When planning an activity which involves the processing of agricultural raw materials, especially short-term perishable products as is the case with fruits and vegetables, accurate planning must take place, to avoid coordination problems. This will prevent significant raw material losses, which often happens in home processing and small industrial scale systems.

Planning is necessary even when very small processing systems are involved, in which the raw material is practically supplied by own production sources. For example, it is possible to conceive a production program based on a large number of families that obtain their products from very small family-run orchards, and then process them in their own home-made installations. It is also possible to produce a given product on a joint basis, so that the production of different centers or families may be sold together and enjoy the benefits of joint marketing. In this case, it becomes crucial to provide a uniform product, to use high-quality raw material in all cases, and naturally, to exactly know the potential volumes to be produced, so as to efficiently determine the marketing mechanism. Such a mechanism begins with planning the production of the raw material.

Production of Raw Material and Processing Techniques

The relationship between raw materials and processing techniques encompasses a series of aspects which range from the selection of a given variety or cultivar of a species, to post-harvest handling and the preservation of the quality of the material to be processed.

In this connection, it should be considered that every processed product requires a specific raw material, if excellent quality levels are to be achieved. There is no reason why a product processed at the level of cottage industries or on a small industrial scale should be of a lower quality than one processed on an industrial scale. To the contrary, home or small scale processing techniques should give better results, as the process is looked after more specifically, the individual units are better controlled, and the raw material may be handled appropriately.

Varieties or Cultivars to Use

There are many different choices possible within a species, for there are varieties or cultivars that present significant differences in terms of their intrinsic characteristics and nature.

To develop a sound industrialization or processing system, it is important to choose the material that presents the best specific characteristics in relation to the purpose of the procedure.

This means that several characteristics of the final product will depend upon the nature of the raw material. For example, a good tomato sauce can only be obtained by using very red tomatoes, at the right stage of maturation, with a firm pulp to ensure a good texture, and with an appropriate content of solids. If the final product is mango nectar, it is preferable to use varieties with little or no fibre, of a bright colour, not astringent and sweet. For pickled cucumbers, it is better to use varieties with a smooth skin, of an intense green colour, cylinder-shaped and with a firm texture.

Therefore, every product will require a raw material that will meet the minimum quality standards that guarantee its suitability for the market. The characteristics vary at times, when consumer have very peculiar preferences when it comes to a given product.

Cultivation, Harvest and Post-harvest Care

The way in which a certain raw material is cultivated, harvest and post-harvest care will determine the material's quality.

All of these aspects are especially easy to take care of through small-scale production systems, where the goods are handled almost exclusively by hand.

However, it is important to consider that especially in such cases, the site of primary production and the processing facility must be close to each other. Indeed, appropriate harvest or post-harvest handling normally cannot be achieved for small volumes when the distance between the orchard and the processing site is great.

When the harvests are small in volume and the distance is great, to save on transportation costs one must resort to initially storing the goods on the production site, but this has a detrimental effect on the quality of the material. If the processing plant is adjacent to the site of production, then all of the problems related to temporary storage are solved, and the material may be processed more rapidly after the harvest.

Finally, it is advisable to adopt a production rationale based on the production of the species presenting comparative advantages in terms of nutritional value, greater demand by potential consumers, and a higher market value, if the products are to be marketed outside the community of origin. This especially applies to raw materials with a greater production cost and for unknown or exotic products.

Production Costs and Processing

An industrial production process normally gives an added value to the product. When using a raw material of great value, the added value will be proportionally lower to that of a material whose quality was poorer at the beginning and which acquires a much greater value as a finished product.

It is therefore preferable for the value of a product to increase as a result of processing, as generally speaking, the product will generate a greater demand and will be more widely accepted.

It is not profitable to use costly processes for very inexpensive raw materials, except for the cases in which the demand is great. An example of this would be frozen peas, where the raw material is relatively cheap and the processing procedure is one of the most expensive, but large-volume production amply justifies this approach.

In species of a greater value, the use of expensive technologies may be justified, as the process only accounts for a small part of the total cost of the product whose raw material is quite valuable.

Agricultural Production Capacity and Processed Output

Every time an industrial production process is planned, whether it be on a small, medium or large industrial scale, there must be some consistency between the potential supply of raw material at the plant and the capacity of the installations to be set up.

Of the two extreme cases that may occur, excessive supply and a shortage of supplies, the former is more difficult to cope with by home production systems. The only way to change the capacity of a small home-processing plant, which is mainly based on manual procedures, is by increasing the number of staff. But this solution is complicated if the workers are not trained to meet the same production, productivity and quality standards as the rest of the permanent staff.

If a temporary or unexpected shortage of supplies occurs, on the other hand, the problem may be solved by working on alternative jobs like labelling, packaging, cleaning the installations, or engaging in other useful tasks. Obviously, devoting excessive time to such jobs may prove to be very expensive in the long run, but at least it allows for some readjustment in times of emergency.

Appropriate planning is required in a small industrial scale plant, for any adjustments weigh more heavily on profitability than they do in a cottage industry. A shortage of supplies may cause serious problems because the installations remain idle, but an excess of supplies may determine even greater problems due to the inflexible nature of the machines.

References

Amoriggi, G. 1990. *Technical Report on Small-Scale Processing of Agricultural Products.* FAO. Strengthening of Agricultural Marketing Service GUY/86/003. Rome, Italy. 23 p.

FAO. 1988. *Rural Processing and Preserving Techniques for Fruits and Vegetables.* Rome, Italy. 59 p.

Holdsworth,: S.D. 1983. *The Preservation of Fruit and Vegetable Food Products.* MacMillan Press, London. 159 p.

Pantastico, ER. B. 1975. *Postharvest Physiology, Handling and Utilization of Tropical and Subtropical Fruits and Vegetables.* The Avi Publishing Co. Inc. Westport, Connecticut, USA. 560 p.

UNIFEM. 1988. Fruit and Vegetable Processing. *Food Cycle Technology Source Book* No. 2. Photosystem SRL, Rome, Italy. 67 p.

3

Raw Material for Processing

Raw materials are one of the most important aspects to consider in fruit and vegetable processing. The fruit and vegetables themselves are the raw materials, the reason for the development of preservation processes. Due to the great number of species suitable for industrialization, only a few will be mentioned, and of these, greater emphasis will be placed on the ones most commonly used.

General Principles

When speaking about raw materials, especially those used by industrial firms and particularly cottage industries, it must be considered that they may have two different origins: they may either grow spontaneously or be cultivated.

In both cases, the quality of the raw material is crucial to the fulfilment of the goals pursued in the processing and preservation of the product, and also determines the level of profit. The material must therefore be of good quality and its industrial performance, which is strongly dependent on the quality of the raw material, must be high. In addition to this, the raw material must meet certain basic sanitary quality requirements.

Production Systems and their Influence on Processing

As stated previously, the quality of a processed product essentially depends on the quality of the raw material. On the other hand, the quality of the raw material also depends on the way that it is handled during the production process.

This is partly true in the case of species that grow in the wild. It is partly true because harvest and post-harvest handling also influences the quality of a product. This is the case of species that are highly sensitive to post-harvest handling like berries, for instance.

However, it is not only the harvest and post-harvest processes that have an impact on the quality of the raw material. The entire production process is important, from planting or sowing to harvesting. And even before sowing, the selection of the soil, of the genetic material to be planted and of the geographical location, will undoubtedly have a significant impact on the final outcome, on the quality of the raw material, and on the processed product.

Of course, some species and specific cultivars or varieties within them, are highly susceptible to environmental conditions, while others are much more resistant to the conditions of the ecosystem in which they grow.

Some of the vitally important factors in the handling of crops or natural resources are presented in the following paragraphs:

— Utilization of cultivars or varieties suitable to the characteristics of the specific environment.

— Technical management of the levels of fertilization required for an appropriate plant growth, reconciling yield with a number of quality-related factors which depend on soil nutrient levels and on the plant. For instance, a proper balance between soil nitrogen and phosphorus, in many vegetables will determine the quality of their colour, texture, and development, and their preservation capacity in the post-harvest stage.

— The control of the plant's water resources is a factor which largely determines final quality. A material that has been somewhat deprived of water will not be suitable for processing. Its sugar and organic acid levels will not allow for healthy development.

— The management of all aspects related to plant health is crucially important in the case of a raw material that must meet minimum quality requirements to be processed, as health standards will determine final quality. For example, certain products intended for dehydration present very serious defects when processed from a fungi infested raw material. Plant health becomes a basic priority in post-harvest preservation. This is a very important aspect of processing in small-capacity cottage industries, where part of the harvested material must often be stored with no refrigeration for a short period of time.

Harvest and Post-harvest Care as a Quality Factor

Harvest and post-harvest care is an aspect of paramount importance, as fruits and vegetables are usually rapidly perishable. Therefore, since industrial performance depends on post-harvest quality, special care must be taken in the period between harvesting of the material and the beginning of processing.

The harvesting method employed and the duration of the harvesting period will also influence the quality of the raw material. Hand harvesting obviously seems more suitable for small plots of land, as those which will be the object of the activities of a small enterprise or a home processing system. In this case, care must be taken to make sure that the harvest operation is performed properly, at the right time and in a way that will not affect the product.

The transportation of the material to the farm and its preservation, the use of containers that will not spoil the material, and transportation from the farm to the plant are factors that also influence the quality of the material to be processed. Very sensitive materials, with a high respiration rate must be processed rapidly or must be stored at relatively low temperatures. Less sensitive materials, on the other hand, do not require such care. Pulses, for instance, must be harvested, transported and subjected to processing very rapidly, for they tend to ripen very quickly.

The post-harvest of such raw materials must be strictly controlled, for they belong to rapidly perishable species. Of course, the idea is to process high quality material, but it is also important to process the greatest amount possible of harvested material. Processing is an alternative way to preserve these products so rich in extremely valuable nutrients, like vitamins, minerals and fibres. Processing must therefore be placed at the service of the preservation of materials that is normally lost in great amounts for want of care.

Fruits

The basic characteristics of a number of fruits suitable for processing and their most important processes are illustrated in the following paragraphs.

Temperate Climate Fruits

These species grow in areas characterized by a temperate climate, that is, where temperatures are never extremely cold.

They include the following species which have an actual and potential economic importance.

Species	*Scientific Latin Name*
Apple	Pyrus malus
Pear	Pyrus communis
Peach	Prunus persicae
Plum	Prunus domestica
Curuba	Passiflora mallisima
Passion-fruit	Passiflora lingularis
Blackberry	Rubus Glaucus
Cherimoya	Annona cherimola
Feijoa	Feijoa selowiana
Strawberry	Fragaria x Annannassa
Tree tomato	Cyphomandra betacea

Tropical and subtropical fruits

Tropical and subtropical fruits include members of the Anacardiaceae family, which comprises about 59 genera and 400 species. Such species are generally found in tropical areas and in high temperature zones throughout the world, as in the Caribbean, Brazil, Central America and Africa. Some plants are considered to be of economic importance, such as mango (Mangifera indica L.), pistachio (Pistacia vera L.) and cashew nut (Anacardium occidentale L.).

These fruits are generally very fragile and sensitive, and therefore require special handling and proper storage conditions. Nevertheless, they are in great demand all over the world and sell at rather high prices, mainly due to the fact that very few countries offer appropriate conditions for their cultivation.

These fruits may be broken down in two groups:

— Fruit trees growing in warm climates of a short, medium and long growth period, of which the following are of great economic importance today.

Species	*Scientific Latin Name*
Mango	Mangifera indica L.
Guava	Psidium guajava
Pineapple	Ananas comosus
Papaya	Carica papaya
Coconut	Cocos nucifera
Lulo	Solanum nucifera
Maracuya	Passiflora edulis

— Fruit trees growing in warm climates of a short, medium and long growth period, and of potential economic importance:

Species	*Scientific Latin Name*
Mangosteen	Garcinia sp.
Carambola	Averhoa carambola
Tamarind	Tamarindus indica L.
Sapote	Matisia cordata
Guava (Coronilla)	Psidium araca
Soursop	Annona muricata L.
Breadfruit	Artcarpus altilis
Tacay or Inchi	Cardiodendron sp.
Caimaròn or uvilla	Pouroma sp.
Borojò	Borojoa patinoi

In the following paragraphs, some general background information will be provided for some fruits with an industrial potential.

Banana

Only a very small part of the banana production is preserved by means of drying, freezing and canning, although such preservation processes are quite important in the areas where these fruits are grown extensively.

The most common industrial forms are dehydrated bananas and banana flour. The latter is produced from fully developed green bananas. Such products are mostly manufactured in Ecuador, Brazil and Costa Rica, among other Latin American countries. Most banana flour is processed with drum dryers, as with spray drying great quantities of product are lost because the material sticks to the equipment. It should be noted that while banana products like flour and pancakes are processed from a well developed and green raw material, in the case of dehydrated bananas the raw material must be ripe.

The varieties most used for dehydrated and dried products are Gros michel, Cavendish, Lady finger and Plantain.

Citrus fruits

The industrialization of citrus fruits requires a raw material with a uniform shape and size. The varieties with a thin and sufficiently hard rind are preferred, as those with a soft rind, like mandarins, require special handling in the preparation and juice extraction processes.

The products obtained from citrus fruits include orange juice in the concentrated and frozen form, and by-products like essential orange oil, washed pulp juice, frozen concentrate, concentrate for animals and d-limonene.

For juice processing, it is essential to use varieties with a high juice content and a good Brix°-acidity balance. The colour is an especially important quality standard in concentrated orange juices, and in the preparation of citrus product bases. Juices squeezed at different times are usually combined to obtain a product with a balanced colour and taste.

Since vitamin C is the most important nutrient in citrus fruit juice, it should be present in great concentrations in the form of ascorbic acid. Another processing requirement is that the raw material must not have an excessively bitter flavour, or that it does not acquire a bitter flavour as a result of thermal processing.

Citrus fruit segments are another derived product. When citrus fruit segments are packaged, it is extremely important for the raw material to have a firm and seedless texture, as deseeding is a very time-consuming and expensive process which spoils the fruits, making them less attractive to consumers.

Grapefruit, mandarin and orange segments are in greatest demand. These fruits must preferably be fully ripe.

Fig

Figs are very suitable for being canned, dehydrated, presented as paste and processed into frozen products or into compote. In their fresh state, however, they spoil rapidly. They are difficult to transport and cannot adapt to storage conditions, not even if they are refrigerated. Figs intended for dehydration must be left to fall from the tree when they are ripe. They must be picked from the ground frequently, to prevent hardening of the skin, the growth of fungi and attacks by insects.

Tree tomato

This fruit is characterized by the fact that it has an ovoid-apiculate shape, that it is green in colour when unripe and reddish-orange when ripe. It is between 6 and 9 cm long, and its widest part measures between 4 and 6 cm. Its average weight may range between 70 and 80 grams. Its skin is thin, smooth and resistant, the pulp has a very pleasant and peculiar flavour, and many seeds are concentrated in the middle of the fruit. This fruit is currently used to make home-made compotes, to make drinks by homogenizing the pulp with water and sugar, to produce spicy sauces, and as seasoning to prepare certain dishes.

Carambola

The carambola is also known as star-fruit; it also has other specific names, according to geographic location. It originated in Ceylon and in the Moluccas, and has been grown in Asia for a very long time. It may be propagated in tropical and subtropical climates and is grown in Australia, the Philippines and other islands of the South Pacific, Central America, South America, the islands of the Caribbean, Africa, Israel and subtropical areas of the United States.

The carambola tree is relatively small and is 6 to 9 m high, its crown being between 6 and 10 m wide. Its leaves are dark green, its flowers have a colour ranging between pink and purple, and a diameter of 6 mm.

The shape of a carambola is between oblong and ellipsoidal, it is between 6 and 15 cm long, with 4 to 6 longitudinal sections, so that when it is cut into cross sections, the fruit has the shape of a star. Its skin is translucent, soft and waxy, and its colour ranges from white to a deep golden yellow. Its taste varies between sweet and sour. The carambola is used to produce juice, nectar, pulp and jam. It may also be preserved in syrup, after having been cut in cross sections.

Lulo (Naranjilla)

The lulo or naranjilla, as it is also known, grows better in the humid valleys of the Andes near Ecuador, at an altitude between 1,200 and 2,100 m. In Ecuador, where it originates, the species is found throughout the country, from the Colombian border to the south, in the Loja province. In Colombia, the main production area is located between Cali and Ipiales.

The fruits are round or slightly oval-shaped, of an orange-yellow colour, with a short peduncle of sepals similar to those of the tomato, adhering to the fruit. As a result of the orange colour and the smooth and resistant aspect of the skin, as well as the predominantly sour taste of the pulp, resembling that of an unripe orange, the fruit is commonly known as "naranjilla" (little orange).

The fruits weigh between 40 and 70 g, and their diameter is comprised between 4 and 5 cm. The internal part of the fruit resembles that of a tomato. The pulp is juicy, of a greenish color, and it is divided in four almost symmetrical sections. The seeds are smooth and roundish, with a diameter of 3 mm and light yellow in colour.

By industrially processing this fruit, the following products may be obtained: nectars and juices, frozen pulps, 65 Brix° concentrate, jams and jellies.

Blackberry

Numerous species of blackberries or brambleberries have been found to grow in the high areas of tropical America, especially in Ecuador, Colombia, Panama, Central American countries and Mexico.

The Rubus and Rosa genera, which belong to the Rosaceae family, are very similar, which is why the blackberry plant rather resembles wild rose plants, with thorns and composite leaves behind five leaflets. The differences between these genera lies in the fruit, as blackberries look like oblong or thimble-shaped strawberries, and when they are ripe they acquire a black, red or purple color. There are believed to be as many as 300 species of blackberries of a relative importance throughout the world, according to the commercial value that they are attributed in the different areas.

The following industrialized products may be obtained from this fruit: nectars and juices, frozen pulps, 65 Brix° concentrate, jams and jellies, 33° Brix concentrates, wine and sulfite pulps.

Cashew

Cashew trees originate from the southern tropical regions like Mexico, Peru, Brazil, as well as Eastern India. However, they are grown in the tropical regions of America, Asia and Africa.

The cashew tree has an average size and can reach a height of 12 m. The fruit (apple) is rhomboidal in shape and is between 5 and 20 cm long, 4-8 cm wide, with a slight red, yellow, or red and yellow skin, which is thin and waxy. The pulp is soft, juicy, yellow, astringent and sour.

Cashews grow better in tropical climates at an altitude below 100 m. They tolerate different degrees of exposure to sunlight, but cannot withstand the cold or floods.

Cashew apples are rapidly perishable. However, the people of India and Latin America consume them in their fresh state and also process them into juices, wines and syrups.

Soursop and Cherimoya

These fruits are rapidly perishable, and must be hand harvested when completely ripe, to prevent them from falling from the tree branches and bruising. The ripe fruit is washed with chlorinated water to remove the soil and minimize the presence of bacteria. Once it is washed, the fruit is peeled and the pits are removed by hand, for there is no current alternative to this procedure.

Soursops and cherimoyas may be consumed as a dessert, although they are mostly used in the form of frozen pulp in foods like ice cream and syrups, and in drinks.

The diluted pulp may be used to produce nectars and juices with special characteristics. The frozen puree is sold with added sugar, up to 59° Brix. Storage life may be prolonged by adding ascorbic acid in concentrations of up to 10-30 g/100 kg. Other products include the combination of soursop pulp and refined tamarind puree and sugar cane or papaya juice.

Guava

The characteristics of a few varieties of this species have been established. Guava is an excellent source of ascorbic acid (vitamin C), and to a lesser degree, of vitamin A, phosphorus, pantotenic acid and B complex vitamins. Its seeds may find a potential application in the production of pectin and oils.

The fruits must have a nice colour, a pH near 3.4 and a solids content between 9 and 12%. The market requires large completely ripe fruits with a firm pulp. Their industrial potential derives from the fact that they may be used to make pulps, purees, powder for nectars, jams, jellies and sweets of 70-75 Brix° (ate).

Mango

Like many other tropical fruits, during thermal processing mangoes undergo chemical changes in terms of their nutritional and organoleptic properties, mainly flavour. It is therefore important to employ procedures that will not affect such thermolabile compounds to a significant degree, like freezing or carefully performed thermal techniques, even at a home-processing level.

Mangoes may be processed into different products, such as puree, frozen pulp, nectar, concentrated and frozen pulp and in a high-sugar pulp preparation known as "ate". Mango pulp may also be dehydrated to produce bars. Mango slices in syrup or in the dehydrated form are also consumed. This fruit is also excellent when pickled.

Papaya

In addition to being widely consumed as fresh fruit, papayas have other applications as food products.

Like other tropical fruits, papayas are prepared and preserved according to different methods. Nectars or juices may be produced by using papaya puree, which either alone or in combination with different-flavoured fruits makes a very tasty product. Papaya pulp is also a very popular product.

Tamarind

This species belongs to the leguminosae family, and every part of the tamarind tree, namely the wood, bark, leaves and fruits, may be used in many different ways. The tamarind has been utilized as medicine since ancient times, for its pulp can combat scurvy and has laxative properties, while its leaves have diuretic properties. However, the tamarind is mostly used as food. The seeds, the soft leaves and the flowers of fully grown trees are utilized in salads and to make soups. Unripe and tender husks are used as seasoning in boiled rice, fish and meats.

The pulp obtained from a ripe fruit is an agroindustrial product of considerable economic value in many parts of the world.

The pulp of the fruit is slightly difficult to extract due to its low water content and, because it is sticky. To remove it, the fruit is normally subjected to a steam bath for several hours. A syrup of about 13.2° Brix may thus be obtained.

VEGETABLES

Vegetables may be divided into the following categories, on the basis of their capacity to adapt to different climates:

Warm Temperate Climate Vegetables

Group A: Vegetables that adapt well to temperatures ranging between 18 and 27°C. They do not tolerate frost. This group includes sweet corn, beans, lima bean, tomato, bell pepper, squash, cucumber and melon.

Group B: Crops characterized by a long growth period, which thrive in areas where the temperature is above 21°C. This group includes the watermelon, sweet potato, eggplant and chili pepper.

Group C: Tropical species growing in very humid areas where the temperature is high. This group includes the "bilimbi".

Cold Climate Vegetables

Group D: These vegetables thrive in areas where the mean monthly temperatures range between 15 and 18°C. They are intolerant to temperatures between 21 and 24°C, and tolerate weak frosts. This group includes spinach, lettuce, broccoli, beetroot, Brussels sprout, cabbage, radish, rhubarb and watercress.

Group E: These vegetables thrive in areas where the temperature ranges between 15 and 16°C. They do not tolerate temperatures between 21 and 24°C. They may be damaged by frost when approaching maturity. This group includes the cauliflower, artichoke, lettuce, green pea, white potato, celery, carrot, chicory, endive, parsley and chard.

Group F: These vegetables adapt well in areas where the temperature ranges between 13 and 19°C. They are tolerant to frost. This group includes the onion, asparagus, garlic, leek and shallot. Some of the vegetables suitable for industrial processing are presented in the following paragraphs.

Tomato (Lycopersicon lycopersicum)

The tomato is a plurannual or perennial plant cultivated as an annual, which belongs to the Solanaceae family.

The fruit is a berry with 2 to 9 loculi. The fruits with many loculi generally have an irregular shape. Their weight ranges between 40 and 300 g.

The tomato is one of the vegetables that enjoys the greatest consumption, both fresh and as a preserve, paste, juice and dehydrated product.

Bilimbi

The bilimbi may be found throughout South-eastern Asia and across Malaysia, which is its region of origin. However, the bilimbi is only known as a species crop. The bilimbi was introduced in Australia, the Caribbean, South and Central America, Florida and Hawaii.

The bilimbi tree can reach a height of up to 18 metres, although it is usually 15 metres tall or less. The bilimbi fruit is cylinder-shaped, between 5 and 7.5 cm long, although its length is usually 5 cm. It is yellow-greenish when ripe, and its skin is thin and soft. Its pulp is green, soft, juicy and very sour, with few seeds.

The bilimbi adapts better to heat, grows well in humid tropical areas, and cannot tolerate freezing temperatures. Young trees may be damaged by temperatures between -1 and -2°C.

The bilimbi is not an important product on the world market as a fresh fruit, although it is processed into jellies, sauces, pickles and - juices.

Eggplant (Solanum melongena)

Eggplant is native to India. It is a perennial solanaceous plant, but is cultivated as an annual.

Its fruit is spherical, elongated and pear-shaped or cylindrical. Its colour is purple when it is ripe, due to the presence of anthocyanin. Some varieties have white fruits. Its pulp is cream coloured and crumbly.

Eggplant may be used for the preparation of preserves in brine and oil, in frozen vegetable mixes, and pickled.

Bell pepper - Chili pepper (Capsicum spp)

There are two types of pepper, the sweet and the hot pepper, the former being used more extensively.

It is a perennial solanaceous plant, grown as an annual. The fruit comprises a composed pericarp, an endocarp and seeds. Inside, it is divided into lobuli. The shape and size of the fruit varies according to the different varieties.

The colour of the fruit is determined by lycopersicine and carotene, and the yellow by the xanthophyll. The scent is determined by its ethereal acid content. The fruit also contains carotene or provitamin A.

Its pungency (degree of spiciness) is determined by the alkaloid called capsicine, whose content ranges from slight traces to 0.71%, mostly concentrated in the pulp.

The product may be consumed directly, or in the form of preserves, pickled peppers or powder.

Carrot (Daucus carota)

The carrot is an umbelliferous biannual root. It is rich in calcium, phosphorus, iron and carotene.

This root is mainly used to prepare pickles and dehydrated products, and is employed in soup mixes, preserves and frozen products, either alone or in combination with other foods. Carrots may be consumed cooked, in syrup and in the form of juice and jam.

Green pea (Pisum sativum)

The green pea originated in Ethiopia and in Mediterranean Europe. It belongs to the Fabaceae family (ex Leguminoseae). It is a climbing, herbaceous annual and requires an amount of water equivalent to 45 cm.

Horticultural green peas may be consumed directly, or as preserves or frozen products. In agriculturally advanced countries, preserved and frozen peas are increasingly taking the market share of fresh peas. Dried grains are used to prepare pre-cooked flours and soups.

Dry bean (Phaseolus vulgaris)

The dry bean is native to Mexico, Peru and Bolivia and originates from the Phaseolus aborigeneus.

It is a leguminous plant, which reaches between 105 and 120 cm, and is distributed extensively on the superficial layer.

The fruit is a pod composed by a pericarp and seeds. For green beans, it is preferable to avoid the formation of parchment between the fleshy parts of the pericarp. The formation of "strings" or "fibres" in the seams of the pod should also be avoided. The dried fruit may be used to prepare soups and dried vegetable mixtures, it may be pickled or employed in the preparation of acidified preserves.

Horticultural beans are consumed directly, both in the green pod as grains and therefore half-ripe, or as preserves. They may also be consumed in the frozen form. Green beans have a low-calorie content, and a high nutritional value because they contain vitamins, minerals and carbohydrates.

Onion (Allium cepa)

Onions are bulbous vegetables, important both in terms of domestic consumption and export. The bulb is consumed in its tender state, raw, ripe, pickled or in form of powder. At a nutritional level, it stimulates appetite even if its calorie content is normal, its protein and dry material content is low, and its vitamin level is not very high.

The bulbs may be red, white or yellow in colour.

Leek (Allium porrum)

The leek originated in the Near East. It does not form bulbs. It is consumed on a small scale in soups, contains less volatile oils than garlic and onion, and is rich in organic sulphur. It is mainly used as a dehydrated product in the preparation of sauces and soups.

Garlic (Allium sativum)

Garlic is native to Southern Europe and Central Asia. It is an annual, of the Amarylidaceas family. The bulb is characterized by antiseptic, diuretic, expectorating, anti-scurvy and antirheumatic properties. It is consumed directly, in the dehydrated form, and may be used in the preparation of pickles and sauces.

Asparagus (Asparragus officinalis)

The asparagus is a perennial vegetable, which contains a high level of thyamine, riboflavine and ascorbic acid. It is consumed in the direct form or as a preserved product. It is native to Europe, the Caucasus and Siberia.

The asparagus belongs to the Liliaceas family. It is made up by a mass of rhizomes with buds on the tip, which give origin to the edible spears. The stalk or turions have a diameter between 6 and 23 mm and grow from buds on the rhizomes. This vegetable is especially used for making dehydrated products and soups.

Artichoke (Cynara scolymus)

This crop is indigenous to Southern Europe. Although it is considered to be a perennial plant, it actually is not, for after it blossoms it dies and is replaced by a shoot. The artichoke belongs to the Compuestas family.

Artichokes are consumed cooked, and the edible portion is the base of the bracts and the heart. Such parts may also be preserved.

Parsley (Petroselium crispum hortense)

Parsley is used as seasoning in soups and sauces. It is a biennial plant and belongs to the family of Umbeliferas.

Coriander (Coriandrum sativum)

This crop is native to southern Europe. It belongs to the family of Umbreliferae. Its leaves are used for seasoning, and its seeds are used to make liquor.

Sweet basil (Ocicum basicilicum)

The green or dried leaves of this plant are used as aromatic flavouring. The crop belongs to the Labiadae family.

Cabbage (Brassica oleracea var. capitata var. subauda)

Cabbage is a very important vegetable due to its high yield. It is used directly in soups or stews and is processed into a fermented product. The crop is native to Asia Minor and the eastern Mediterranean region.

Cauliflower (Brassica oleracea var. botrytis)

Cauliflower may be consumed directly or pickled. It is indigenous to the island of Cyprus. It may be used to prepare pickles, alone or in combination with other vegetables.

Cucumber (Cucumis sativus)

Cucumber is consumed directly in salads and pickled. It is not very important from a nutritional point of view, as it has a 95-96% water content and few vitamins. It originated in the humid areas of India, and was later exported to China and South-eastern Asia. Cucumber may be used to make dehydrated products, it may be preserved in oil or vinegar or be processed into naturally fermented pickles.

Pumpkin (Cucurbita spp)

Pumpkin is important in terms of its composition and of its relatively high yield. It may be consumed directly in stews and in pies or in syrup. It is native to Central and South America. The plants are annuals and bear large fruit that may even weigh 50 kg. Their pulp has a variable thickness and may be white, off-white, yellow, etc. in colour. The pulp may be used to make pies or it may be preserved as acidified and sterilized pieces in syrup.

Zapallo hoyo and zucchini (Cucurbita spp)

The so-called zapallos tiernos include hoyo, zucchini and hoyito. They may be consumed when they are still unripe in soups or stuffed. Their vitamin and sugar content is lower than in that of pumpkins.

These vegetables have an elongated shape, their surface is rough and they are dark green in colour. Zucchini may be preserved as dehydrated products or in oil, after having been sterilized.

The great diversity of raw materials existing both in this region and in the rest of the world, is such that it would be impossible to engage in an exhaustive analysis of all of the species suitable for industrialization.

In the specific case of vegetables, practically all species may potentially be subjected to industrialization processes, with the exception of lettuce, which can only be fermented. The rest of the species, however, are all suitable for processing.

In the case of fruit, the number of both tropical and subtropical species that may potentially be subjected to processing greatly exceeds that of species not suitable for industrialization.

AUXILIARY RAW MATERIALS

Auxiliary raw materials used in fruit and vegetable processing technologies play a major role in the determination of their physical and chemical characteristics, sensory properties and nutritive value.

Water

Water is one of the essential factors in the activity of the processing centres; according to the final utilisation, water can be classified in three categories:

a. for technological utilisation (when it comes into direct contact with raw materials and enters in the finished product's composition),
b. for steam generators and
c. for receptacle cooling, washing of equipment and general hygiene.

Water for technological uses

Water coming into direct contact with the raw materials used for processing (washing, blanching, etc.) or that used as filling liquid of some canned products, must be of drinking water quality in terms of its physico-chemical and microbiological conditions.

More important even than fulfilling drinking water standards, water used for these purposes pests certain specific characteristics related to the technological step or the raw material treated during the processing.

When very hard water is used for blanching vegetables some pecto-calcium and pectomagnesium complexes are formed which starts the hardening of vegetable tissues. This process continues over the pasteurization of the finished product.

When fruit is processed in sugar syrup, the use of hard water for the syrup preparation could induce the formation of a pectin-sugar-acid gel facilitated by the medium pH and presence of calcium salts.

Soft water has negative consequences associated with mineral and hydrosoluble substances and losses during blanching of vegetables.

For some specific products such as peeled tomatoes, green beans and fine texture fruit, the addition of calcium salts (chlorure, sulphate, etc.) is employed to correct for low texture.

The water hardness is an essential factor when used as filling liquid for canned products; ideally, the hardness of the water should be adapted to the raw material species used for canning.

Thus a hardness of 3° is good for beans, 5° to 9° for green peas, green beans, and for fruit and vegetables with a tendency to disintegrate should use even harder water.

In the technological process of cucumber and gherkin preservation by natural acidification (lactic fermentation), water hardness has a paramount role. The literature maintains that as far as texture is concerned, the best results are obtained by using very hard water (about 30°); but since high magnesium and salt content has a negative effect on taste, in practice it is recommended that a water hardness of about 10° be used, which gives satisfactory results from both points of view.

Oxygen present in water can act as a corrosion factor in metal receptacles but this negative influence can be eliminated by preliminary boiling.

An important factor is pH. Water for canning must be neutral or slightly basic. Acid water plays a major role in corrosion which is evident both on receptacles and on iron or copper equipment, where changes of product colour will be induced. More dangerous is the attack on lead pipes or to the mix used for can sticking and this can render the product toxic. For these reasons an acid water must be neutralised before use.

Water for steam generators.

Two main conditions must be fulfilled:

- hardness has to be as low as possible, even zero, because precipitation of calcium salts can lead to the formation of encrustations ("crusts") in pipes and on equipment walls. For this reason, water treatment is practiced in the majority of installations;
- from a bacteriological point of view, the iron-bacteria must be eliminated with biological filters or oxidising substances. This is necessary in order to avoid iron hydroxide formed during equipment exploitation to deposit on the inner walls of the pipes.

The elimination of iron-bacteria is also of importance for the water used in processing steps.

Water for receptacle cooling and general hygiene.

This should be of drinking water standard. Where this is not available in sufficient quantity, the use of industrial water is acceptable but only for cleaning of production rooms/workshops.

Sweeteners

Sugar

Sugar is the conventional name applied to sucrose. Physically there is icing, granulated and lump sugar. In fruit and vegetable processing, sugar is used only in its granulated form; this quality must be in the form of uniform crystals, white, shining and completely soluble in water.

Concentration of various sugar solutions can be rapidly measured by refractometer reading or with areometers graduated in various ways: Brix, Baumé, etc.

Sugar solubility in water is dependent upon temperature; for example, in order to obtain a saturated solution, one must dissolve 2040 g in one litre of water at 20° C and 4870 g at 100° C. Taking into account this temperature related solubility, in practice the majority of sugar solutions are prepared by heating the water. Water should be as soft as possible because the calcium salts can precipitate on boiling.

Corn syrup (liquid glucose)

Corn syrup is obtained industrially by acid or enzymatic starch hydrolysis, using as starting raw materials maize (corn) or potatoes. In fruit processing, mainly in the production of marmalades, it is possible to use corn syrup. The average composition of this corn syrup is of about 32-40% dextrose (glucose), about 40% dextrins and 18-20% moisture. Sweetening power is 50% compared with sucrose.

In a 10%-20% proportion with sucrose, addition of corn syrup has certain advantages:

a. it improves the shine and texture of marmalade;

b. it prevents "sugaring" defect and

c. it reduces the too sweet taste of finished products obtained with sugar alone.

Salt

Salt is used in order to give to the finished products a specifically salty taste and as a preserving substance. From a chemical point of view the term salt means sodium chloride but in practice the product is never in a pure state. The presence of a significant quantity of magnesium chloride increases the hygroscopicity, gives a bitter taste and can induce corrosion of receptacles.

Table 1. Physical characteristics of sugar solutions

Specific weight	*deg.Bé*	*Sugar in solution*		*Boiling temperature °C*
		deg.Bx (K/100g)	*g/l*	
1.144	18.5	33	377	101.1
1.149	19.0	34	391	101
1.154	19.6	35	404	101.2
1.159	20.1	36	417	101.3
1.164	20.7	37	430	101.3
1.169	21.2	38	444	101.4
1.174	21.8	39	457	101.4
1.179	22.3	40	470	101.5
1.185	22.9	41	486	101.5

1.190	23.4	42	500	101.6
1.195	23.9	43	513	101.6
1.200	24.5	44	527	101.7
1.206	25.0	45	543	101.7
1.211	25.6	46	557	101.8
1.216	26.1	47	571	101.8
1.222	26.6	48	586	101.9
1.227	27.2	49	600	101.9
1.233	27.7	50	616	102.0
1.238	28.2	51	630	102.1
1.244	28.8	52	646	102.2
1.249	29.3	53	660	102.3
1.255	29.8	54	677	102.4
1.261	30.4	55	693	102.5
1.267	30.9	56	709	102.6
1.272	31.4	57	723	102.7
1.278	31.9	58	740	102.8
1.284	32.5	59	757	102.9
1.290	33.0	60	774	103.0
1.296	33.5	61	790	103.2
1.302	34.0	62	807	103.5
1.308	34.5	63	824	103.7
1.314	35.1	64	840	103.9
1.320	35.6	65	857	104.2
1.326	36.1	66	874	104.6
1.332	36.6	67	891	105.0
1.338	37.1	68	909	105.5
1.345	37.6	69	928	106.0
1.351	38.1	70	945	106.5
1.357	38.6	71	962	106.8
1.364	39.1	72	981	107.1
1.370	39.6	73	999	107.4
1 376	40.1	74	1017	107.8
1.383	40.6	75	1037	108.2
1.389	41.1	76	1058	109.0
1.396	41.6	77	1076	110.0
1.402	42.1	78	1094	111.0
1.409	42.6	79	1113	112.0
1.414	43.1	80	1133	113.0

From a microbiological point of view, salt it is not a sterile product but on the contrary contains various micro-organisms, mainly halophil bacteria.

Salt solubility is only slightly influenced by temperature (0.360 kg/1 at 20° C and 0.390 kg/l at 100° C). Correspondence between specific weight and salt content of salt solutions at 15° C is shown in Table 2.

Table 2. Physical characteristics of salt solutions

deg.Bé	*Specific weight*	*NaCl content*	
		g/100 g	*g/l*
1	1.007	1	10
2	1.014	2	20
3	1.022	3	30
4	1.029	4	41
5	1.037	5	52
6	1.045	6	63
7	1.052	7	74
8	1.060	8	85
9	1.067	9	96
10	1.075	10	107
11	1.083	11	119
12	1.091	12	131
13	1.100	13	143
14	1.108	14.2	158
15	1.116	15.5	173
16	1.125	16.7	188
17	1.134	18.0	204
18	1.142	19.0	217
19	1.152	20.0	230
20	1.162	21.2	246
21	1.171	22.4	262
22	1.180	23.6	278
23	1.190	24.8	295
24	1.200	26.0	312
24.5	1.204	26.4	318

Food Acids

Acetic acid

Acetic acid is in use as solutions of various concentrations which are known under the generic name of vinegar. Vinegar can be obtained:

a) from wine, alcohol, cider, beer, etc. by fermentation;

b) by dilution of acetic acid obtained by dry wood distillation or by synthesis.

From a quality point of view, wine vinegar is preferred, as it has a more pleasant flavour. In order to improve taste, other vinegar types are usually flavoured with spices.

In addition to its spicing and flavouring role, vinegar is used and acts as a preservation agent for some vegetables: cucumbers, acidified vegetables, etc.

— Citric acid

— Tartric acid

— Lactic acid

Pectic Preparations

In fruit processing there many preparations and mixes known as "pectin" are used as liquid or powder extracts.

From a practical point of view, the pectic preparations are classified as:

— strong pectins obtained from apples or citrus fruit peel; this category gives gels rich in sugar;

— weak pectins which gives gels with low proportion of sugar or even without sugar but with the addition of calcium salts.

The fruit industry uses mainly strong pectins. These preparations are characterised, from a commercial point of view, by the capacity of gelification, expressed in degrees. The degree of gelification represents the quantity of sugar in grams able to be transformed in a standard gel (65% sugar and pH=3) by 1 g pectin.

Intensive Sweeteners

"Calorie-reduced" and "low-calorie foods are widely used and are cornering an increasing share of the market. Sweeteners are making an important contribution to the manufacture of sweet foods in these categories. They make it possible to manufacture sweet products without "sweet" being necessarily synonymous with "high-calorie".

Diabetics need to restrict their intake of sugar and various carbohydrates similar to sugar or avoid them altogether. Sweeteners enable diabetics to enjoy sweet tastes without changing their lifestyle. Sweeteners do not contribute to the development of tooth decay; they do not degrade in the mouth to form the acids which are responsible for caries. Thus sweeteners offer consumers a number of advantageous and favourable properties above and beyond merely reducing calories.

Sunett is the trade mark of Hoechst AG for its high intensity sweetener acesulfame K. As an ingredient, it can be used for sweetening all foods produced industrially or at home, or to produce tabletop sweeteners. Like all other sweeteners, Sunett tastes more intensely sweet than sugar; it is about 200 times sweeter than sugar compared with a dilute stock solution.

Synergism. Sunett is notable for its pronounced synergism with other sweeteners. The synergistic effect leads as quantitative synergism to an intensification of the overall sweetness and as qualitative synergism to an improved taste. The synergism results in a marked intensification of the sweet taste of the blends, which can amount to 30-50% at usual concentrations.

The favourable properties of Sunett, particularly its synergistic behaviour, can be used advantageously in the Sunett-Multi-Sweetener Concept. For high sweetness levels, blends of Sunett with other sweeteners, e.g. aspartame, are particularly favourable in many applications. The synergistic effect of the blend results in a taste that is particularly pleasant and rounded.

Alone or in blends with other sweeteners, Sunett is used mainly where only sweetness is important and no other properties of the foods are to be affected. This applies particularly to beverages (fruit syrups and juices, carbonated beverages, etc.).

In a number of food products, sweet carbohydrates do not only provide the sweet taste; they have other functions to fulfil; for example, they act as bulking or texturing agents and as preservatives by reducing the water activity. In these types of products, Sunett and other sweeteners cannot be used on their own. They must be combined with other substances which perform the required functions. These may be bulking agents or sugar substitutes (for example polyols: sorbitol, mannitol, xylitol). Sunett can be combined with both groups of substances.

Uses. In carbonated soft drinks blends of Sunett with other sweet-tasting substances are often recommended. Beverages formulated using the Sunett Multi-Sweetener Concept, i.e. blends of Sunett with other sweeteners, are usually preferred because of their particularly well balanced and rounded sweetness profile. The taste of these blends is often superiors to single sweeteners.

Sunett is compatible with sugar and other sweet-tasting carbohydrates, both technically and in terms of taste. It can therefore be used in the production of soft drinks with a reduced sugar content. Drinks based on fructose and Sunett are suitable for diabetics. Soft drinks with a reduced sugar content show an excellent taste quality when Sunett or Sunett blends are added to bring them up to the usual level of sweetness.

Fruit nectars and fruit juice drinks differ from most carbonated drinks in that they contain a noticeable amount of sweet carbohydrates provided by the fruit juice. The amount of carbohydrate may vary depending on the type of fruit and amount of juice used.

Amounts of up to 200 mg/l Sunett are often adequate as a single sweetener for the popular types of fruit nectars. With blends of Sunett and aspartame quantities in the order of 100150 mg/l Sunett and 50 mg/l aspartame or 60-70 mg/l of both Sunett and aspartame are sufficient.

Sunett offers such excellent stability that end products, such as drinks, show no reduction in sweetness performance during normal processing methods and storage periods. Sunett can withstand pasteurization, hot filling and aseptic filling without any loss of sweetness.

Jams and marmalades. Sugar contributes a great deal to the texture and stability of conventional jams and marmalades. For the production of sugar-free products either the sugar must be replaced by comparable amounts of sugar substitutes or some of the functions of the sugar must be taken only by other components, such as suitable gelling agents.

Sugar-free jams and marmalades containing sweeteners are more susceptible to microorganisms than sugar-containing products. The risk of spoilage due to yeast fermentation or moulding can be prevented by pasteurization. However, this is only feasible for small jars which will be quickly consumed once they are opened. In all other cases, it is advisable to add 0.05-0.1% potassium sorbate as a preservative, wherever this is permitted under the relevant food regulations.

For sugar-free jams and marmalades, concentrations in the range of 500-2000 mg Sunett/kg of the finished product are appropriate. It is advisable to add Sunett in the form of an aqueous stock solution towards the end of the boiling process. Care must be taken to ensure that Sunett is evenly dispersed throughout the whole batch.

Because of the excellent compatibility of Sunett with sugar alcohols, fruit jams and marmalades using these ingredients offer an outstanding taste.

Fruit preserves. Sunett can be used for the production of sugar-free or sugar-reduced fruit preserves. At the normal pH values for fruit preserves Sunett can be added even before pasteurization, as the sweetness is not impaired under the usual thermal treatment conditions. Sunett also withstands the usual storage periods without loss of sweetness.

NutraSweet(r) is the commercial name of aspartame (APM), a new sweetener from G.D. SEARLE & Co. which can be used in the most foods in order to give the same taste as sugar. NutraSweet is about 180 to 200 times sweeter than sucrose (sugar) and this value depends on pH, temperature and the type of flavour.

NutraSweet can be used as mentioned above in a mix with Sunett or alone in all sugar-free or calorie-reduced fruit jams, marmalades and preserves. Like other sweeteners, NutraSweet does not promote tooth decay.

International regulatory status. Both Sunett and NutraSweet are widely accepted by food laws in the majority of countries. The following are the main fruit products where Sunett is an accepted sweetener:

— low-joule prepared jelly: max. 500 mg/kg;
— canned fruit without added sugar: max. 500 mg/kg;
— beverages including calorie-reduced fruit nectars: max. 600 mg/l;
— calorie-reduced jams;
— calorie-reduced and dietetic fruit compotes;
— canned fruit and vegetables, fruit puree, jams and marmalades;
— juices, nectars and juice based beverages;
— jams, marmalades and related products: max. 300 mg/kg.

REFERENCES

FAO. 1990. Preserving Tomatoes. *Three Home-processing and Preservation Techniques.* TCR/BKF/6658, Rome, Italy.

Fellows, P. and Hampton, Ann. 1992. *Small-scale Food Processing. A guide to appropriate equipment.* Intermediate Technology Publications. London, UK. 158 p.

Mc Williams, Margaret and Paine, Harriet. 1977. *Modern Food Preservation.* Plycon Press, California, USA. 198 p.

Nagy, S., Shaw, P. and Wardowski, W. 1990. *Fruits of Tropical and Subtropical Origin.* Florida Science Source Inc., USA. 391 p.

4

Processing of Fresh-cut Fruits and Vegetables

Fresh-cut products are fruits or vegetables that have been trimmed, peeled and/or cut into a fully usable product, which is subsequently packaged to offer consumers high nutrition, convenience and flavour while maintaining freshness.

The market for chilled fresh-cut produce has witnessed dramatic growth in recent years, stimulated largely by consumer demand for fresh, healthy, convenient and additive-free foods which are safe and nutritious. The food industry has responded to this demand with creative product development, new production practices, innovative use of technology and skillful marketing initiatives.

Fresh-cut tropical fruits on the market today include melons, cantaloupe, watermelon, mangoes, mangosteen, rambutan, jackfruit, pummelo, papaya, durian, grapefruit, pineapples and fruit mixes. Fresh-cut salads on the market include shredded leafy vegetables and salad mixes. Fresh-cut vegetables for cooking include peeled baby carrots, baby corn, broccoli and cauliflower florets, cut celery stalks, shredded cabbage, cut asparagus, stir-fry mixes and cut sweet potatoes. Fresh-cut herbs are also marketed widely.

Growing consumer interest in international markets in new or exotic tastes has promulgated growth in the international trade of fresh-cut products. Tropical fresh-cut fruits are particularly attractive to young consumers and 'baby boomers' who consume these products as snacks. Manufacturers in many tropical countries have responded to this growing demand by producing fresh-cuts for export.

Consumers generally purchase fresh-cut produce for convenience, freshness, nutrition, safety and the eating experience. Consumer demand for these attributes has, indeed, led to considerable innovation and diversification in the fresh-cut industry. Apart from presenting the consumer with a range of options in a single package, fresh-cuts reduce wastage at the household level, in that they allow the consumer to procure only the quantities of fresh produce required, while allowing the opportunity to readily assess the quality of the produce being purchased.

Whilst the production of fresh-cut produce requires relatively little product transformation, it requires investment in technology, equipment, management systems and strict observance of food safety principles and practices to ensure product quality.

Trends in Fresh-cut Market

Healthy foods are of critical importance to the baby boomer population in the United States' market. With shifting population demographics, consumption of ethnic foods also continues to show a growth trend. A range of tropical fruits and vegetables is imported into the United States to ensure that the year-round demand for fresh produce is adequately met.

With increased dining away from home, fresh-cut products are playing an ever-increasing role in the United States' food service sector. In 2006, 27 percent of fresh-cut produce in the United States was sold in the food service sector, while 73 percent was sold in retail. Fresh-cut produce sales increased in value from US$3.3 billion in 1999 to US$15.5 billion in 2007. Bagged salads and cut vegetables showed a growth trend in 2008, while sales in fresh-cut fruits declined. Fresh-cut organic salads are now being mainstreamed across the United States and feeding consumer desire for healthy food and preservation of the environment. They are now widely available in restaurants and retail outlets.

A major trend in the United States' fresh-cut industry over the past two decades has been the consolidation of companies at all levels of the food chain. Food service and retail buyers have formed conglomerates around the world and food processors have also consolidated to keep up with this trend. Large processors are now supplying large retail chains, which creates tough competition for small producers. Some regional companies have consolidated to form larger companies in order to supply growing food service chains. Small locally-owned or family-owned fresh-cut processing facilities can still fill a niche where flexibility in product type and small quantities are needed for regional restaurants and grocery stores. Small processing facilities may produce minor quantities of a range of fresh-cut products in one day to fill regional needs.

The European fresh-cut industry has shown exponential growth since the early 1980s. According to Rabobank consumer emphasis on convenience and healthy living is the key driver for growth in the fresh-cut fruits and vegetables sector, with retail as the major distribution channel. While the Netherlands, Switzerland, Italy and Spain have already established a market and show strong growth in the fresh-cut produce sector, Germany has yet to embrace the sector. Italy is currently emerging as one of Europe's leading fresh-cut markets, where sales of fresh-cut fruits and vegetables have been booming in recent years and are now second only to the United Kingdom in terms of value.

Fresh-cut produce is sold in open-air markets and food stands in many Asian countries and is increasingly being sold in supermarkets. Fresh-cut fruits, in particular, have gained popularity in urban centres of the region. Often these products are displayed without the benefits of refrigeration so their shelf-life is frequently not extended beyond the day of display.

Fresh-cut vegetables command a larger market share than fresh-cut fruits in Thailand. With growing consumer demand for ready-to-eat products, the market for fresh-cut products in Thailand is likely to show a continued growth trend.

The market for fresh-cut products in Japan and Republic of Korea has shown a steady growth trend since the late 1980s and 1990s respectively. While the food service industry for school meals and restaurants is the main user of fresh-cut products in these countries, demand for them has grown in retail markets. Fresh-cut vegetables for cooking constitute the largest part of the fresh-cut produce industry in both countries. Fresh-cut salads are another major category as consumers perceive them to be healthy. Fresh-cut fruits continue to show a rapid growth trend in these countries. However, with increasing demand for fresh-cuts at the retail level, the fresh-cut industry in Japan and Republic of Korea is facing challenges to extend shelf-life and enhance food safety.

The production of traditional dishes in most developing countries necessitates a variety of fresh ingredients. The drudgery of peeling vegetables, shelling peas and trimming herbs and vegetables, and then combining these ingredients, often deters the busy housewife from preparing these traditional foods. Similarly the difficulty of peeling fruits such as pineapple, durian and pummelo and sometimes their large size deters the consumer from purchasing them. Fresh-cut processing addresses all of these issues by making products available in a convenient, easy-to-use format with minimal waste. Packs of fresh-cut fruits and vegetables are increasingly being prepared by cottage industry suppliers in many developing countries and are being sold in wet markets in response to consumer demand for produce in a ready-to-use format.

Cottage industries and small vendors are still the major supplier of fresh-cut fruits and vegetables in most developing countries. Food caterers often supply fresh-cut fruit packs for school feeding programmes. Supermarkets in most developing countries produce fresh-cut fruits and non-leafy fresh-cut vegetables on site to meet consumer demand. Bagged leafy vegetables are supplied primarily by small enterprises engaged in fresh-cut production. Growth in the fast food sector and in the food service industry of many developing countries has also increased market opportunities for many small producers of fresh-cuts.

Developed countries are constantly looking for innovative products, providing potential for tropical fresh-cut fruits and vegetables to fill a gap. Food products offering new colours, flavours and textures, if creatively packaged, would be welcome additions to the fresh-cut industry in developed countries.

Equally profitable would be the sale of fresh-cut tropical fruits and vegetables in the urban centres of developing countries. The convenience of preparing traditional meals from locally-grown ingredients would preserve culinary traditions, promote the consumption of local produce and allow consumers to eat the foods they enjoy. As the fresh-cut fruit and vegetable industry has been so profitable in developed countries, it is anticipated that the same level of growth would take place in urban centres of developing countries enjoying income growth, whose citizens have less time for meal preparation.

Developing countries have experienced many demographic changes in the past two decades. With population growth, urban centres are expanding on all continents. At the same time, traditional food supply chains and food habits have been changing to keep up with these changing trends. These social changes include:

— Increases in single person households;
— Increases in middle-income populations;
— Less time for meal preparation;
— Increased demand for convenience food items;
— Increased sales of ready-to-eat meals; and
— Increases in restaurant and fast food operations.

Growth in the market opportunities for fresh-cut produce will continue, only if consumers believe that fresh-cut produce is safe and of high quality with sufficient shelf-life. Other challenges to the marketing of tropical fresh-cut produce include:

— Preserving product quality through the marketing chain;
— The inherent fragile nature of some fruit;
— Maintaining the cold chain and proper logistics;
— Adequacy of processing equipment, refrigerated storage and processing facilities;
— Availability of technology to set up processing plants and conduct research to preserve the quality of tropical fresh-cut produce.

Tropical fresh-cut fruits and vegetables face significant challenges in entering export markets, owing to:

— Failure to meet permitted pesticide residue levels;
— Allowable levels of microbial contamination;
— Food safety standards and audit standards;
— Food laws of the importing country;
— Quality standards of the buyer (private standards); and
— Compliance with international regulations such as the Codex Alimentarius of the World Health Organization (WHO) and the Food and Agriculture Organization of the United Nations (FAO).

Compliance with these regulatory requirements is necessary in order to develop a viable export industry for fresh-cut tropical produce from developing countries. Certification programmes that guarantee the quality and safety of fresh-cut produce are critical to their entry in export markets.

As many tropical fruits and vegetables are less well known in developed countries, a systematic approach is needed to promote them as 'novel' products that must taste and look appealing to the consumer in developed countries.

Factors that influence the marketing success of novel tropical fresh-cut products include:

— Meeting customer needs for quality and taste;
— Meeting customer expectations for price;
— Product presentation and consumer appeal;
— Ability to respond to current social food values - organic, fair trade, health options; and
— Satisfying the labeling requirements in the country of sale.

There is a need to provide as much information as possible about the supplier of the raw material in order to promote sales in those sectors of society wanting to make a difference in developing countries.

Tropical Fruits

Tropical fruits come in a variety of shapes and sizes, with a range of specific flavours and textures. They are typically consumed when in season in tropical countries. Consumers often prefer sun-ripened fruits, with fully developed flavours.

Mangoes, papayas, pineapples, lychees, bananas and various varieties of melons have been exported by developing countries for decades. Developing countries account for 98 percent of tropical fruit production. Mangoes account for 38 percent of tropical fruit output globally, papayas account for 14 percent and avocadoes account for 4 percent. Pineapples remain popular at about 21 percent of the output. The so-called minor tropical fruits now account for almost 25 percent of global output and include lychees, durian, rambutan, guava and passion fruit (granadilla).

According to FAO's *Tropical fruits commodity notes,* while the dramatic increase in production of tropical fruits is now levelling off, the growth trend in the export of these fruits continues. The growing diaspora of foreign nationals in many developed countries has fuelled growth in tropical fruit consumption in these countries. For example the fruit import centre in Birmingham, United Kingdom, has been almost solely driven and managed by immigrant traders. A growing population, interest in things exotic and a trend towards healthier eating, has increased the demand for tropical fruits and vegetables across many European countries. Many popular tropical fruits require peeling and preparation for eating. They are attractive to the busy consumer if presented in a convenient package in a ready-to-eat format.

Carambola *(Averrhoa carambola)* or five-finger star fruit, is a species of tree native to India, Indonesia and Sri Lanka. It is popular throughout the Caribbean, parts of East Asia, Southeast Asia (for example Malaysia) and in many other tropical countries. Popularly known as star fruit, carambola is crunchy, and has a slightly tart, acidic, sweet taste, reminiscent of pears and apples.

Guava derived from the Arawak 'guayabo' via the Spanish 'guayaba' is a genus of about 100 species of tropical shrubs and small trees in the Myrtaceae family of myrtles; it is native to Mexico, the Caribbean, Central America and northern South America. It has a thin delicate rind which is pale green to yellow in colour at maturity in some species or pink to red in others. It has a creamy white or orange-salmon flesh with many small hard seeds. Guava has a strong, characteristic aroma.

Jackfruit is a species of the mulberry family (Moraceae) and native to Bangladesh, southwestern India, the Philippines and Sri Lanka. It is the largest tree-borne fruit in the world. The fruits can reach 36 kilograms in weight and up to 90 centimetres in length and 50 centimetres in diameter. The sweet yellow sheaths around the seeds are about 3-5 millimetres thick and have a taste similar to that of the pineapple but are somewhat milder and less juicy.

Mangoes are native to South and Southeast Asia and account for almost 50 percent of all tropical fruits produced worldwide. Mango production in 2004 was estimated at 26.6 million tonnes and ranked seventh in world fruit production behind bananas, grapes, oranges, apples, coconuts and plantains. The top ten mango-producing countries in 2007, based on area of production, were China, Brazil, Guinea, India, Indonesia, Mexico, Nigeria, Pakistan, the Philippines and Thailand. The top five mango-exporting countries were Brazil, India, Mexico, Peru and the Philippines. Bangladesh, the Netherlands, Saudi Arabia, the United Arab Emirates and the United States were the top five mango-importing countries.

Mangoes contain essential vitamins and dietary minerals. The antioxidant vitamins A, C and E comprise 25, 76 and 9 percent of the dietary reference intake (DRI) in a 165-gram serving of mango. Mangoes also contain other nutrients such as B vitamins and essential nutrients such as potassium, copper and amino acids at reasonably good levels. Mango peel and pulp contain other phytonutrients, such as pigment antioxidants – carotenoids and polyphenols – and omega-3 and -6 polyunsaturated fatty acids.

Mangosteen *(Garcinia mangostana)* is a tropical fruit, believed to have originated in the Sunda Islands and the Moluccas. This fruit is either dark red or purple in colour. Its thick skin houses a sweet white pulp when fully ripe. Botanically an aril, the fragrant edible flesh of the mangosteen can be described as having a sweet and tangy, citrus-like flavour and texture. A subtle mild aroma, with fewer constituents than other fragrant fruits, explains this relative mildness.

Papaya, fruit of the plant of the genus Carica, is the most economically important fruit of the Caricaceae family. Native to southern Mexico and neighbouring Central America, the papaya is now widely cultivated in tropical countries such as Brazil, Haiti, India, South Africa and Sri Lanka as well as in the Philippines and elsewhere in the Southeast Asian region. According to FAOSTAT, Brazil was by far the largest producer of papaya globally in 2007, followed by India, Indonesia, Mexico and Nigeria.

The ripe papaya fruit, with its delicate flavour, is a tasty addition to fresh-cut fruit salads or served on its own as a chilled dessert. The ripe fruit is generally eaten in the fresh form, without the peel or seeds. The unripe 'green' fruit of papaya can be eaten in the cooked form in curries and stews, or as a salad. It is also rich in pectin, which makes it suitable for use in the production of fruit jellies.

Papaya fruit are very delicate and soften when ripe. The leaves are the source of the meat-tenderizing enzyme, papain which can be used in marinades for tenderizing tough meat cuts.

Pineapple fruit are found in most tropical and subtropical countries. The pineapple has its origins in Southern Brazil and Paraguay. Popular pineapple varieties such as Smooth Cayenne, Formosa, Red Spanish and Carbezona have been commercialized for decades. According to FAOSTAT, Brazil, Costa Rica, Indonesia, the Philippines and Thailand were the five leading producers of pineapples globally in 2005.

Pineapples are the only *bromeliad* fruit in widespread cultivation. The pineapple has multiple, helically-arranged flowers, each producing a fleshy fruit pressed against the adjacent fruits, forming what appears to be a single fleshy fruit. Pineapple is eaten in the fresh form or is preserved by canning and/or juicing. It is used in desserts, salads, as a complement to meat dishes and in fruit cocktails. Although it is sweet, it also has a high fruit acid content.

The low-acid hybrid, Cayenne is the most common fresh pineapple variety sold in the United States and in European supermarkets. This variety was developed in Hawaii in the 1970s. Fresh pineapple is delicate and thus poses a challenge during shipment. Pineapples will ripen after harvest, but like bananas are chilling-sensitive and are not readily stored under refrigerated conditions.

The pummelo is a citrus fruit native to Southeast Asia. It is usually pale green to yellow when ripe, with sweet white (or, more rarely, pink or red) flesh and very thick rind. It is the largest citrus fruit, 15-25 cm in diameter, and usually weighing 1-2 kg.

Rambutan is a medium-sized tropical tree of the family *Sapindaceae.* The fruit of this tree is probably native to Southeast Asia and is closely related to the lychee. The leathery skin of this fruit is reddish (rarely orange or yellow) in colour and is covered with fleshy green pliable spines. The fruit flesh is translucent, whitish or very pale pink, with a sweet, mildly acidic flavour. The single seed is glossy brown, 2-3 centimetres long, with a white basal scar. The seed is soft and crunchy, is edible and may be, but is not usually eaten with the pulp.

Sugar apple or atis is thought to have its origins in the Caribbean. Sugar apple is also referred to as *sweetsop*. The fruit has a scaly skin, is almost round and about 10 centimetres in width. The flesh of the ripe fruit is slightly sweet and is soft in texture. Many black pips are scattered through the flesh of the fruit.

Tropical Vegetables

Vegetables generally require a great deal of preparation prior to their inclusion as ingredients in recipes for soups, stews and ethnic dishes in some countries.

Bitter melon is a tropical and subtropical vine of the family Cucurbitaceae, also referred to as corolla or cerassie. It is among the most bitter of all vegetables. Throughout Asia, the leaves and fruit of this plant are used in soups, stir fries and teas. The original home of the species is not known, other than that it is a native of the tropics. It is widely grown in Africa, China, the Caribbean and South and Southeast Asia.

Pak choi or Chinese cabbage, also known as snow cabbage, is a leafy vegetable often used in Asian cuisine. This vegetable is related to the western cabbage and is of the same species as the common turnip. It is one of the most well-known vegetables and is also one of the oldest green vegetables in Asia. The leaves are green, mild flavoured and less crisp than other cabbages. It is often used in stews and soups.Chayote whose common names include Cho-cho, or Chouchou, is a yellow-green or white, pear-shaped vegetable. It has a wrinkled scabby skin, with greenish-white flesh. The chayote is believed to have its origins in Central America. More common varieties of this vegetable are pear- or apple-shaped, and are somewhat flattened with coarse wrinkles, ranging from 10 to 20 centimetres in length. The flesh is fairly bland in taste, and has a texture described as being a cross between that of a potato and a cucumber.

Pumpkin *(Cucurbita maxima)* or squash is a vegetable that is cultivated throughout Southeast Asia; it can be cooked with or without the skin. The colour of the skin can vary from green to yellow. The flesh is deep yellow to orange in colour. Pumpkins are soft in texture and are sweet. They are rich in vitamin A and are consumed as side dishes or as ingredients in bakery products.

Edible stems

Ginger *(Zingiber officinale)* is a monocotyledonous perennial plant. The term ginger is also used to describe the edible part of the plant which is commonly used as a spice in cooking throughout the world. Erroneously referred to as 'ginger root', the edible section is actually the horizontal subterranean stem or rhizome of the plant. The ginger plant has a long history of cultivation known to have originated in China; it later spread to India, the Caribbean, Southeast Asia and West Africa.

In 2005, China continued to lead the world in ginger production with a global share of almost 25 percent followed by India, Indonesia and Nepal.

Heart of palm *(palmito, pajibaye)* also referred to as palm heart, palmito, chonta, or swamp cabbage, is a vegetable harvested from the inner core and growing bud of certain palm trees (coconut, palmito, juçara

[*Euterpe edulis*], açaí palm [*Euterpe oleracea] etc.*). It is costly because harvesting in the wild kills the tree. Heart of palm is often eaten in salads. It is sometimes referred to as the 'millionaire's salad' and is also used in vegetarian spreads.

When harvesting the cultivated young palm, the tree is cut and the bark is removed leaving layers of white fibres around the centre core. During processing the fibres are removed leaving the centre core or heart of palm. The centre core is attached to a slightly more fibrous cylindrical base with a larger diameter. The entire cylindrical centre core and the attached base are edible. The centre core is considered more of a delicacy because of its lower fibre content.

Sugar cane is a genus *(Saccharum)* of up to 37 species, depending on the taxonomy, of tall perennial grasses (family Poaceae, tribe Andropogoneae) that are native to warm temperate to tropical regions. They have stout, jointed, fibrous stalks that are rich in sugar and measure 2 to 6 metres in height. The edible portion of sugar cane is the inner stalk (stem) the sap of which is a source of sugar. The thick stalk stores energy as sucrose in the sap. The sap can be extracted by pressing the sugar cane stalk.

From this juice, crystallized sugar is extracted by evaporating the water. As of 2005, the world's largest producer of sugar cane by far was Brazil. Uses of sugar cane include the production of crystal and liquid sugar, molasses, rum and ethanol for fuel.

Fresh Produce Quality And Safety

Quality is a combination of characteristics that determines the value of produce to the consumer. Consumers expect fresh produce to be without defects, of optimum maturity and in fresh condition. The condition of fresh fruits and vegetables relates to their general appearance, sensory quality, flavour and nutritional quality. Consumers judge the quality of fresh fruits and vegetables based on their appearance and firmness (external attributes). Subsequent purchases by consumers are, however, dependent on their eating experience (aroma, taste, texture intrinsic or experience attributes). Other quality parameters like nutritional quality and safety (hidden attributes) increasingly influence consumer decisions on making repeated purchases of the commodity. Only fruits of the highest quality should be used for fresh-cut processing.

Components of Fresh Produce Quality

Quality attributes are often classified as external, internal or hidden. External quality attributes are visual when the product is first encountered. These attributes are generally related to 'appearance' and 'feel'. Internal quality characteristics are sensed when the product is cut or bitten. Internal attributes include aroma, taste and feel (for example, 'mouth-feel' and 'toughness'). Hidden quality attributes include wholesomeness, nutritional value and safety of the product.

Appearance quality

The visual quality of fruits and vegetables relates to their size, shape, colour, glossiness, cleanliness of the surface and absence of defects or signs of spoilage. Fresh produce defects include evidence of: bruising and crushing of pieces, shrivelling and wilting due to water loss, mushiness or tissue softening, colour changes due to enzymatic browning or physical disorders, sliminess and water soaking due to ageing. Spoilage also induces changes in appearance, such as mushiness owing to mould growth and package swelling due to the release of gases by bacteria or unattractive colour changes.

Texture quality

Texture *(feel)* relates to the feel of the produce item in the hands or in the mouth. Textural attributes can be described in the context of firmness or hardness, crunchiness, crispness, tenderness, juiciness, mealiness and toughness, depending on the commodity. The textural quality of horticultural produce is not only important for its eating and cooking quality, but also for its transportability through the horticultural chain.

Flavour quality

Flavour (eating) relates to the taste and smell (aroma) of the produce. Flavour attributes can be described as the perception of tastes and aromas coming from many sources including sugars for sweetness, the acidity of organic acids such as citric acid in oranges and lemons, the bitterness and astringency of phenolic compounds and tannins, and aromas from volatile compounds. Most fruits and vegetables have sweet, sour and bitter tastes but little or no salty and umami1 tastes.

Good flavour is an important guarantee for repeat purchasing of fruits by consumers. Consumers are generally willing to pay more for desirable flavour. Growth of spoilage microorganisms can lead to off-odour or flavour development, as occurs when fresh produce begins to ferment.

Eating quality

The eating quality of fresh produce results from complex interactions between texture and flavour and effects on the nutritional value.

Nutritional quality

Fruits and vegetables are well-known sources of useful nutrients in the form of vitamins, minerals, dietary fibre and other phytonutrients including flavonoids, carotenoids and phenolic compounds that may lower the risk of cancer, heart disease and others illnesses. Consumption of fresh fruits and vegetables is encouraged and a recommendation of at least five servings per day has been made to American consumers.

Safety

Safe food is free of physical and chemical hazards or microorganisms that cause adverse effects to human health and life. Safety is a component of quality. Many experts are of the view that safety is the most important component of quality, since unsafe food can result in serious injury and even death. Business loss and a poor image as a result of legal action are also serious consequences of consuming unsafe food.

The safety of fresh produce can be compromised by *physical hazards* such as glass, dust and insects or chemical hazards such as pesticides or microbiological hazards resulting from poor sanitation or bad hygienic practice in the producer to consumer chain.

Biological hazards refer to pathogenic microorganisms that cause human illnesses directly when consumed together with the produce (infection) or by producing toxins or chemical substances harmful to humans in the produce before it is consumed. In some cases, the microbial population may be insufficient to cause produce decay but may be enough to cause human infection or intoxication after the produce has been consumed. Thus, produce that is perfect in appearance is not guaranteed to be microbiologically safe. Pathogenic microorganisms can contaminate the produce through the soil, contaminated water, badly treated manure, sewage, air, poor worker hygiene or contaminated surfaces of post-harvest facilities such as produce packages, transport and storage facilities.

Chemical hazards include natural substances (e.g. allergens, mycotoxins, alkaloids and enzyme inhibitors), chemical products (e.g. pesticides, water disinfectants), prohibited substances (e.g. some pesticides, methyl bromide) and toxic elements (e.g. lead, cadmium, arsenic, zinc). Their adverse effects on human health are generally less dramatic and immediate than those caused by pathogenic microorganisms. However, there is growing concern for their possible long-term effects on human health, direct and indirect effects on the environment, flora and fauna, and effects on the health of rural workers. Chemical hazards can be introduced to fresh fruits and vegetables during production (e.g. phytosanitary products, fertilizers, antibiotics, growth regulators, etc.) and post-harvest handling (e.g. phytosanitary products, waxes, detergents, etc.).

Physical hazards include stones, glass particles, wood, hair, plastic, jewelry and metals which may be unintentionally introduced to the produce during production and post-harvest operations.

Evaluation of Quality

Quality evaluation methods can be both destructive and non-destructive. They include objective methods, based on the use of instruments and subjective methods, based on human judgement. Instruments may be designed to imitate human-testing methods. Fundamental mechanical measurements may be statistically related to human perceptions and judgements to predict quality categories. Quality can only be judged by consumers. Instruments that measure quality-related attributes are vital for research and inspection.

An additional dimension of quality which has evolved in international markets, relates to credence attributes, i.e. attributes that depend on the method of production, regardless of whether the method of production has a visible or analysable impact on the produce. Examples of credence attributes desired by consumers include sustainable environmental profiles or fair trade conditions.

Objective measurements of quality employ the use of instruments (e.g. colorimeter for colour, firmness tester for texture) and are suitable for routine quality control but cannot measure consumer preferences. The only way to assess consumer preferences for fruits and vegetables is by sensory testing and asking panellists to provide their opinions/assessment on the fresh produce item. Sensory evaluation is not, however, suitable for routine use and the best way of assuring quality is to find objective measurements that correlate to sensory attributes.

Evaluation of appearance and colour

Colour perception depends on the type and intensity of light, the chemical and physical characteristics of produce and the ability of an individual to perceive the colour. Colour is measured using subjective and objective methods.

Subjective methods of colour measurement rely on the human eye. A colour index is formulated and is matched with a colour chart that corresponds to the developmental stages of a specific produce item. Colour indexing and colour chart matching are, however, prone to human error because of differences in colour perception. Trained panelists may be required to reduce the bias. The light exposure and quality of light at the time of measurement may also affect colour perception.

A colorimeter is used for the objective measurement of colour and can detect small differences in colour. Obtaining colorimetric measurements is however costly and more time consuming than using colour indices.

Evaluation of taste and flavour

Sugar is the main component of total soluble solids (TSS) in most fruits and vegetables. Measurement of TSS, therefore, provides a reasonable indicator of sugar levels or sweetness. TSS is measured using either a refractometer or a hydrometer. The former operates on the basis of the refraction of light by juice samples and the latter on the basis of the density of the juice. Both methods are 'destructive', only needing a few drops of sample juice for measurement.

Acidity is generally measured by titration with a suitable alkaline solution such as sodium hydroxide or by measuring the pH using a pH meter. Certain acids can be individually measured by high performance liquid chromatography (HPLC). Light reflectance in near infrared (NIR) has been applied as a non-destructive method for measuring sugar levels in mangoes, melons and kiwi fruit.

Sensory studies can be used to identify optimal harvest maturity, evaluate flavour quality in breeding programmes, determine optimal storage and handling conditions, assess effects of disinfestations or preconditioning techniques on flavour quality and measure flavour quality over the post-harvest life of a product. Sensory difference testing can be used to measure slight differences in the aromas of produce. Panelists are trained to detect a range of flavour attributes and score their intensities, generally on a 150-millimetre unstructured line.

Laboratory measurements are conducted by headspace analysis using gas chromatography and purge and trap headspace sampling methods that involve trapping and concentrating volatile components on a solid support. Volatiles are later released from the trap using heat for analysis by gas chromatography-mass spectrometry (GC-MS). This method is excellent for quantification and identification of aroma compounds.

Solid phase micro extraction (SPME), a rapid sampling technique wherein volatiles interact with a fibre-coated probe inserted into sample headspace is also applicable in sensory assessment. The probe is transferred to a GC injection port where the volatiles are desorbed. Aside from GC and GC-MS methods, new sensors with a broad range of selectivity are used in headspace analysis. These sensor arrays, referred to as 'electronic noses', are useful in discriminating reactions of volatile components.

Evaluation of texture

According to Meilgaard *et al.*, texture is a consequence of four major properties:

— Surface properties: surface moistness or wetness;

— First bite properties: springiness, hardness, denseness, crispness, juiciness and uniformity of bite;

— Chew down properties: chewiness and cohesiveness; and

— After swollen properties: mouth coating, toughness.

Texture is generally determined on the basis of measurements of force applied to a food sample. The firmness of fresh-cut produce is, in many instances, a fairly good indicator of textural properties. The most common method of texture measurement is the puncture test which is conducted with the use of a penetrometer such as the Magness-Taylor firmness tester or the Effegi penetrometer.

The penetrometer measures the total force required to puncture through a given portion of the fruit or vegetable to a standard depth using a probe of specified diameter. The rate of loading should be controlled and specified in mechanized measurements. The optimal rate of loading differs for different commodities.

Compression tests are not commonly used by the fruit and vegetable industry; they are widely used in research. They can be made on cylindrical tissue specimens excised from the fruit or on intact products using a variety of contact geometries.

Factors that impact on fresh produce quality

The pre-harvest quality of produce is influenced by the cultivars, genotypes and rootstocks, climate, cultural practices, maturity and ripening.

Pre-harvest factors

Cultivar: Genetics are the key to the flavour, texture, post-harvest life and susceptibility of fresh fruits and vegetables to enzymatic browning. The selection of appropriate cultivars is, therefore, of critical importance in assuring the quality characteristics of fruit destined for fresh-cut processing.

Climatic factors: Climatic factors and the environment in which crops are produced have a significant impact on the eating quality and shelf-life of fresh produce. Temperature and light intensity pre-harvest influence the make-up and nutritional quality of fruits and vegetables. The location and growing season may influence the levels of ascorbic acid, carotene, riboflavin etc. High levels of rainfall increase the susceptibility of the plant to mechanical damage. Low light intensity generally results in reduced levels of ascorbic acid (vitamin C) in plant tissue.

Temperature influences the uptake of minerals by plants during transpiration. The strength of the sun's radiation is known to affect produce texture. Excessive exposure of tomatoes to excessive sunlight can, for example, result in sunscald or sunburn. Intense sunlight can increase fruit temperatures and results in fruit damage and loss of firmness.

Cultural practices: Cultural practices such as pruning and trimming enhance crop load and the size of fruit. Fruit maturity may be affected by the use of pesticides and growth regulators. Soil quality with respect to nutrient composition has an immediate impact on developing fruit and on texture, appearance and taste. Kader and Mitcham reported that a lack of calcium during cultivation often resulted in tissue softening after harvest. Bachmann and Earles reported that produce that has been stressed by too much or too little water, or high rates of nitrogen, was particularly susceptible to post-harvest diseases.

The impact of cultural practices on food safety is now widely recognized. A number of outbreaks of foodborne illnesses have been traced to contamination of produce in the field. Improper use of manure and contaminated irrigation water can transfer pathogens onto crops, resulting in human disease. Raw manure should never be applied to edible crops. Good Agricultural Practice (GAP) must therefore be applied in order to mitigate contamination risks to produce during production.

Harvest factors that impact on fresh produce quality

The maturity of fruits and vegetables at harvest determines their quality. Immature fruit may lack flavour and are susceptible to disorders. Apart from affecting the flavour, maturity at harvest also has a direct effect on the texture of fresh produce. Immature fruits contain pectic materials, known as protopectins, of very high molecular weight. On ripening, the chain length of the pectic polymer decreases and water-soluble pectins are formed. This results in textural changes in the fruit, leading to a soft and mushy consistency. Overripe fruits are highly susceptible to damage during cutting and are thus unsuitable for fresh-cut processing.

Many fruits are harvested at the mature but unripe stage of development in order to ensure best eating quality. Climacteric fruits such as bananas, guavas and papayas continue to ripen after harvest. These fruits can be ripened by exposure to ethylene. Other fruits such as pineapples, oranges, lychees and muskmelons must be harvested when mature because no further ripening and flavour development occurs.

Kader and Mitcham suggested that fully mature fruit in which metabolic activity is in decline are at an optimal stage for harvest given that post-harvest changes occur more slowly and storage life is improved. Ripe and fully ripe fruit tend to develop a desirable taste with the onset of starch to sugar conversion, as occurs in bananas and mangoes for example. Astringency due to a high level of tannins also tends to decline at this time. Watada and Li noted that the quality of young honeydew melon cubes with 8.8 percent soluble solids was lower than that of more mature fruits with 13 percent soluble solids after storage. Beaulieu and Baldwin found that the aroma of fresh-cut cantaloupe increased with harvest maturity.

Vegetables are generally of optimal eating quality prior to reaching full maturity. Vegetables that are not harvested at the correct stage of maturity are generally of lower quality and undergo rapid deterioration after harvest.

Post-harvest factors that impact on fresh produce quality and safety

Quality after harvest is influenced by handling and the management of relative humidity and temperature conditions. Quality loss results from factors that are both internal (physiological processes) and external (microbiological, chemical, environmental and mechanical) to harvested produce. Respiration and transpiration are physiological processes that can be greatly influenced by environmental conditions such as temperature, relative humidity, the composition of the gaseous environment and mechanical or physical damage to the produce.

Microbial and chemical contamination can greatly compromise the safety of horticultural produce. Microbial contamination can be transmitted through improper cultural practices, by unsanitary worker practice and through contact with soil and unclean surfaces. Mechanical injury can accelerate the loss of water and vitamin C and can increase susceptibility to decay by pathogenic microorganisms.

Maintaining quality after harvest necessitates proper handling to avoid mechanical and physical injury and the avoidance of chemical and microbiological contamination. Produce must also be stored under optimal conditions of temperature and relative humidity. Ethylene-sensitive commodities such as leafy vegetables, herbs and watermelons must be stored in the absence of ethylene and away from high ethylene producers such as cantaloupes, peaches and tomatoes.

Maintaining Quality of Fresh Produce between Harvest and Processing

Harvesting to assure quality and safety

Quality is influenced by maturity at harvest and the methods of harvesting, storage times, temperatures and extent of handling. Produce should ideally be harvested during the cooler part of the day, on the day of processing, in order to minimize storage and handling and thus increase fresh-cut shelf-life. Field staff should be trained in proper harvesting methods so as to prevent damage to the produce. Fruit should be physically mature, with flavour developing for best taste. Produce with insect damage, sun scorch or any other physical damage would not produce good quality fresh-cut products with the desired shelf-life. Reducing defects also reduces the microbial load associated with the produce.

Field workers must use the appropriate harvesting equipment and protective containers in order to prevent damage to produce during harvesting.

Post-harvest handling to assure quality and safety of fresh produce

Harvested produce should be placed in a shaded area so as to avoid sun damage if awaiting transportation to the processing plant. Produce must be properly handled to avoid bruising and contamination.

Pre-cooling

Pre-cooling is the rapid removal of field heat from fresh produce. It is among the most efficient quality enhancements available to commercial producers and ranks as one of the most essential value-added activities in the horticultural chain. Proper pre-cooling can:

— Prevent quality loss due to softening by suppressing enzymatic degradation and respiratory activity;

— Prevent wilting by slowing or inhibiting water loss;

— Slow the rate of produce decay by slowing or inhibiting microbial growth (fungi and bacteria);

— Reduce the rate of ethylene production; and

— Minimize the impact of ethylene on ethylene-sensitive produce.

Washing and disinfection

Any dirt on the surface of produce must be thoroughly removed by washing in water. The produce must be subsequently washed in potable water containing a sanitizer in order to reduce the risk of the transfer of microbial contamination during processing.

A sanitizing agent or sanitizer is an antimicrobial agent applied to destroy or reduce the number of microorganisms of public health concern, without affecting produce quality and consumer safety. Sanitizers minimize the transmission of pathogens from water to produce, reduce the microbial load on the surface of the produce and prevent microbial build-up in the processing water.

Sanitizers applied to fresh fruits and vegetables must be safe in use and must be used in accordance with given instructions. The sanitizer concentration in the processing water should be routinely monitored and adjusted to prescribed levels. Should it not be possible to monitor the sanitizer concentration, recommendations for the reuse of sanitized water should be followed. In order to minimize the build-up of organic materials, the water must be filtered, or otherwise changed. Technical assistance on the use of sanitizers should be sought when necessary.

Sanitizer treatment: Post-harvest handling and sanitation treatments have considerable impact on the microbiological quality of fresh-cut produce. Washing whole fruits and vegetables in clean water only achieves an insignificant reduction in microbial populations. The use of sanitizers such as chlorine, peroxyacetic acid, hydrogen peroxide, acidified sodium chloride or ozone can provide an additional 1-2 log reduction in the initial population of microorganisms on the surface of fresh produce.

Chlorine is currently the most commonly used sanitizer in washing operations. Chlorine has been successfully used at concentrations ranging from 50 to 200 parts per million (ppm) to wash fruits in fresh-cut fruit studies and fresh-cut mango studies. Improper use of chlorine can, however, lead to the retention of a faint chlorine odour on the fresh-cut fruit.

Despite its common use as a disinfectant for decontamination of fresh produce, chemical hazards associated with chlorine or chlorine residues are of concern. The use of chlorine in the processing of minimally processed products has been banned in some European countries.

Alternatives to chlorine

Chlorine dioxide (ClO_2): This is a water-soluble yellowish green gas with an odour similar to that of chlorine. It does not hydrolyse in water and is unaffected by pH (6-10) changes. It does not react with organic matter to form chloroform. It has been approved for use in flume waters in fruit and vegetable operations by the United States Food and Drug Administration (USFDA). The oxidizing power of chlorine dioxide is 2.5 times that of chlorine. Chlorine dioxide is effective against many microorganisms at lower concentrations than free chlorine. It is highly effective at neutral pH. Its reactivity is, however, reduced by the presence of organic matter.

Chlorine dioxide is more costly than chlorine. It cannot be transported and must be generated on site. In addition, simple assays for routine evaluation of its concentration are not available. Chlorine dioxide may produce hazardous by-products such as chlorite (ClO_2) and chlorate (ClO_3). Its noxious odour is toxic to humans. Microbial susceptibility to chlorine dioxide depends on the microbial strain and environmental conditions during application. Chlorine dioxide can be used on processing equipment at a maximum level of 200 ppm whereas for whole or uncut produce it can be used at a level of 3 ppm.

Acidified sodium chlorite (ASC, $NaClO_2$): This is a chlorine-based sanitizer synonymous with SANOVA® and forms chlorous acid ($NaClO_2$ + H+ – $HClO_2$), which has strong oxidizing capacity. Chlorous acid further breaks down to chlorite. This sanitizer has been approved by the USFDA and the United States Environmental Protection Agency (USEPA) for application on fruits and vegetables, including fresh-cut produce, by spraying or dipping in 500-1 200 ppm solution.

Ozone (O_3): This is a water-soluble gas with broad and rapid biocidal activity. It has strong oxidizing capacity and high reactivity and penetrability. It is, however, unstable under ambient temperature conditions. Ozone rapidly undergoes spontaneous decomposition under conditions of high pH (pH >8) leading to the production of oxygen which is a non-toxic product. Ozone must be generated on site from air.

Gaseous ozone is toxic to humans (>4 ppm). The maximum permissible level for short-term exposure is 0.3 ppm in air. It is corrosive to common materials, thus stainless steel should be used. Ozone must be filtered in order to remove organic and particulate materials. Ozone has been given a generally recognized as safe (GRAS) status for use in food contact applications. Concentrations of 1 ppm or lower in water and short contact times are sufficient to inactivate bacteria, moulds, yeasts, parasites and viruses. An ozone concentration of 0.5-4 ppm is recommended for wash water, whereas for flume water an ozone concentration of 0.1 ppm is recommended.

Electrolyzed water: This may come in the form of acidic electrolyzed water (AEW) or neutral electrolyzed water (NEW). AEW is known as electrolyzed oxidizing water and is strongly acidic (pH of 2.1 to 4.5). It contains HOCl as an antimicrobial component. AEW is used widely in Japan. It has biocidal effects against *E. coli* O157:H7, *S. enteritidis, L. monocytogenes* and biofilms. The bactericidal power of AEW is higher than that of a 5-ppm ozone solution in the decontamination of fresh-cut lettuce.

Fresh-Cut Processing

Fresh-cut processing involves peeling, trimming and deseeding fresh produce and cutting it to specific size. Fresh-cut products must not only look fresh, but must have the sensory properties – aroma, taste, texture and visual appeal – associated with freshly prepared produce. Thus only fresh produce of good quality must be used as the starting material in fresh-cut processing. Fresh-cut products must also be safe, wholesome and nutritious.

Physiological Effects of Fresh-cut Processing

Fresh-cut processing involves cutting through the tissue of fresh produce, thus causing major tissue disruption and the release of enzymes that interact with substrates associated with the fruit tissue. Wounding of the fruit tissue by cutting also increases ethylene production and stimulates respiration and phenolic metabolism. Phenylalanine ammonia lyase (PAL), an enzyme that catalyses the formation of phenolic compounds, is stimulated by ethylene production. Phenolic compounds in turn serve as substrates for polyphenoloxidase enzymes which, in the presence of oxygen, eventually lead to the formation of complex brown polymers.

Increased respiration rates result in water loss and a reduction in the levels of carbohydrates, vitamins and organic acids, with a net negative impact on flavour and aroma. Water loss is also enhanced by membrane and cell wall degradation, and results in loss of turgor. At the same time, microbial growth at the cut surface also increases as sugars become available, thus accelerating the opportunity for microbial spoilage.

Biochemical Changes brought about by Fresh-cut Processing

Colour change

Browning or surface darkening is one of the main physiological effects of fresh-cut processing and leads to quality loss in fresh-cut produce. It is the result of oxidation of phenolic substrates present in the produce by PPO enzymes. The extent of browning is dependent on the concentrations of active PPO and phenolic compounds in the produce tissue, pH, temperature and oxygen available to the tissues as well as on the presence of antioxidant compounds. High levels of PPO enzymes are generally found in tissues that are rich in phenolic compounds. Levels of PPO and PPO substrates change during the life cycle of fruits and vegetables.

Carotenoids, a yellow pigment in fruit and vegetable tissues, are highly susceptible to oxidative breakdown that is catalysed by lipoxygenase enzymes. Yellowing of green vegetables such as broccoli and spinach reduces their quality and shelf-life. Dehydration of surface debris on cut and peeled carrots results in a translucent appearance, referred to as 'white blush', which reduces their market appeal.

Flavour quality changes

Key components of flavour in fresh fruits are sweetness, acidity, astringency and bitterness. Many flavour and aroma components are lost in fresh-cut fruits through enzymatic reactions brought about by cutting, and through the increased respiration rates of the fruit tissue.

Microbial spoilage also contributes to flavour degradation in fresh-cut products. Fresh-cut products can acquire off-flavours with the growth of lactic acid bacteria or *pseudomonads*, resulting in fermentation and the production of acids, alcohols and carbon dioxide gas (CO_2). Lipase enzymes and the breakdown of amino acids in fruits by microorganisms can contribute to the alteration of fruit flavours.

Changes in nutritional quality

Vitamins A, B6, C, thiamine, niacin, minerals and dietary fibre all contribute to the nutritional quality of fresh fruits and vegetables. Compounds such as flavonoids, carotenoids, polyphenols and other phytonutrients associated with reducing the risk of cancer and heart disease are also found in plants. According to Gil *et al.,* fresh-cut fruits and vegetables can appear visually spoiled before any nutrient loss occurs. In the future, plant-breeding techniques may be used to create cultivars with improved nutritional attributes that are able to withstand the effects of processing.

Texture quality changes

The unprotected cut surface of fresh-cut fruits loses moisture at an extremely rapid rate. Such high rates of water loss result in rapid wilting and shrivelling of fresh-cut produce and thus a loss of the crisp, firm texture of the product. Tissue softening of fresh-cut produce during storage is the result of structural changes in the primary cell walls; this is caused by enzymatic activity that leads to dissolution of the rigid pectic cells and a decrease in their resistance to pressure. Decreased rigidity due to water loss is the main cause of tissue softening in fresh-cut fruits.

While there is a paucity of data on the effects of fertilizer on fresh-cut fruit quality, too much nitrogen is known to reduce firmness, while high levels of potassium and calcium can improve fruit quality at harvest.

Quality loss due to microbial contamination

Fresh-cut vegetables harbour lower numbers of microorganisms than unwashed whole vegetables, as a result of washing in chlorinated water. Slicing, dicing and shredding procedures, as well as temperature abuse during storage, can, however, result in increases in populations of mesophilic aerobic microorganisms associated with fresh-cut products. The effects of processing and storage conditions on the survival and growth of pathogenic microorganisms on fresh-cut produce is a public health concern.

Microbes associated with fresh-cut fruit and vegetable products can vary greatly in accordance with the produce type and storage conditions. Temperature plays a significant role in determining the nature of the microflora associated with refrigerated fresh-cut fruits and vegetables. The numbers and kinds of microorganisms associated with fresh-cut produce are highly variable. Mesophilic bacteria from plate count studies typically ranged from 103 to 109 CFU/g. Total counts on products after processing ranged from 103 to 106 CFU/g.

Vegetables are susceptible to attack by bacterial pathogens owing to their neutral pH. Spoilage of fresh-cut vegetables by bacteria is characterized by brown or black discoloration, production of off-odours, loss of texture and soft rot. Fruit products undergo fermentative spoilage by lactic acid bacteria or yeasts and wilting owing to vascular infections.

Spoilage Organisms Associated with Fresh-cut Produce

Pseudomonads: The family Pseudomonadaceae consists of the four genera: *Pseudomonas, Xanthomonas, Zoogloea* and *Frauteuria.* The role of pseudomonads as spoilage organisms is well known. Pseudomonads contribute to the spoilage of fresh produce by producing tissue-degrading enzymes such as pectolytic cellulases, xylanases, glycoside hydrolases and lipoxygenase that degrade the cell walls of the plant tissue, resulting in maceration of the plant tissue. *Pseudomonads* may also contribute to the yellowing of vegetable products during storage, through the production of ethylene.

Pseudomonads are commonly isolated from fresh-cut vegetable salads. *Pseudomonas fluorescens* are generally the main species associated with the spoilage of leafy vegetables such as fresh-cut spinach and lettuce. Other *Pseudomonas* species associated with fresh-cuts include *Pseudomonas putida, Pseudomonas chloraphis, Pseudomonas corrugata, Pseudomonas cepacia, Pseudomonas paucimobillis, Pseudomonas marginalis* (*P. fluorescens* biotype II) and *Pseudomonas viridflava*. Predominant species reported on vegetables such as broccoli, endive and sprouts include the fluorescent *Pseudomonads* and species of *Klebsiella, Serratia, Lavobacterium, Xanthomonas, Chromobacterium* and *Alcaligenes*. *Pseudomonas. Syringae* and *Pseudomonas stutzeri* were isolated from stored fresh-cut Thai mango. These plant pathogens occur on many plant species and can contribute to the spoilage of fresh produce when their populations exceed 108 CFU/g.

Lactic acid bacteria: These are gram-positive bacteria, which can be both rods and cocci. They are traditionally known as fermentative organisms and are associated with fermented food products and with food spoilage. There are three main groups: the *Lactobacillus delbrueckii* group, which includes mainly homofermentative lactobacilli; the *Lactobacillus casei/ Pediococcus* group; and the *Leuconostoc* group. The habitats of species of the genera *Lactobacillus* and *Leuconostoc* include plants and plant material, soil, water, sewage, fruit and grain mashes. Their fermentative metabolism and ability to grow under anaerobic conditions enable lactic acid bacteria to cause spoilage, such as souring of the product, gas production and slime formation. Fresh produce items that are susceptible to spoilage by lactic acid bacteria include lettuce, chicory leaves and carrots.

Yeast and moulds: Fruits are frequently affected by fungal pathogens because of their relatively high levels of acidity (low pH) while vegetables are susceptible to attack by bacterial pathogens because they are neither acidic nor basic, but are at a neutral pH. In some instances, specific plant pathogenic fungi may be associated with particular forms of spoilage such as the breakdown of tissues. Moulds are aerobic microorganisms and can, therefore, be inhibited at CO_2 concentrations as low as 10 percent.

Pathogenic organisms

Food-borne pathogens known to contaminate fresh-cut products include bacteria, viruses and parasites such as protozoa. Of these, bacteria are of the greatest concern in terms of reported cases and gravity of illness. Although not their natural habitat, most fruits and vegetables contain nutrients required to support the growth of pathogenic and toxigenic microorganisms. Storage temperature and pH are reported to be the two principal determinants of growth for food-borne pathogens associated with fresh produce.

Psychrotrophic bacteria, which are organisms that can grow under conditions of refrigeration, vary widely in their acidic tolerances and are the most important spoilage group for fruits and vegetables. The most important of these from a food safety point of view are *Listeria* and *Clostridium*. The fact that these organisms can grow at refrigerated temperatures makes them very important from an export perspective.

Fresh-cut vegetables can be occasionally contaminated with food-borne pathogenic bacteria such as *L. monocytogenes* and *Salmonella* spp..

Good Manufacturing Practice (GMP) and food safety systems such as Hazard Analysis and Critical Control Points (HACCP) must be applied in the production of fresh-cut fruits and vegetables if they are to be recognized as being safe.

Listeria monocytogenes: This is a gram-positive bacterium that causes listeriosis, a serious disease in pregnant women, the elderly and those with weakened immune systems. *L. monocytogenes* is widespread in the environment (i.e. in soil, water and decaying vegetation). It has been isolated from the intestinal tract of domestic animals and humans, from raw produce and in food-processing environments particularly in cool damp areas. It can survive and grow under both ambient and refrigerated temperatures. There are reports describing survival of the organism and formation of biofilms on surfaces in food-processing environments, particularly in drains. It may also be transmitted by aerosols and on workers' hands.

Listeria monocytogenes has been isolated from pre-packaged mixed vegetable products, chicory, endive and fresh-cut lettuce, sliced cucumber and fruits such as tomatoes and cantaloupe. It has also been implicated in food-borne disease outbreaks across the globe.

Salmonella: Salmonella is a common cause of food-borne illness (salmonellosis) and is responsible for millions of cases of illness each year. Fresh produce may become contaminated with salmonellae through contact with sewage and contaminated water or through handling by infected workers. *Salmonellae* do not grow in foods stored at temperatures lower than 7°C.

Salmonella should, therefore, not pose a risk to public health in fresh-cut products, provided these products are maintained at or below 7°C.

Typical symptoms of salmonellosis are nausea, vomiting, abdominal cramps, fever, mild diarrhoea and headache. These symptoms generally last over six to 48 hours.

Clostridium botulinum: This is a spore-forming bacterium. The spores of *C. botulinum* are commonly found in agricultural soils and on the surfaces of fruits and vegetables; they generate potent neurotoxins that produce a range of symptoms in humans including nausea, diarrhoea and vomiting and neurological symptoms such as blurred vision, dilated pupils, paralysis of motor nerves, loss of normal mouth and throat functions, lack of muscle coordination, other complications and possible death. *C. botulinum* spores are capable of growing on fresh-cut vegetables under conditions of low O2 and high temperature.

Shigella spp: Humans are a natural reservoir for *Shigella* spp. The primary means of transmission of the *Shigella* organism is by the fecal-oral route. Most cases of infection by *Shigella* (shigellosis) are attributed to the ingestion of food or water contaminated with fecal matter. Contamination has often been associated with poor personal hygiene of food workers. Typical symptoms include abdominal pain, cramps, diarrhoea, fever, vomiting and blood, pus, or mucus in stools. Shigellosis outbreaks have been associated with shredded lettuce, potato salad, green onions and parsley.

Escherichia coli: It is reasonable to believe that as a result of substandard or even illegal agricultural practices, produce may be contaminated with human pathogens such as *E. coli*. T h e National Advisory Committee on Microbiological Criteria for Foods (NACMCF) in the United States lists 11 agents associated with produce-borne outbreaks. Foremost among them are *Escherichia coli* O157:H7 and various *Salmonella* serotypes.

Control of food-borne pathogens

Control of food-borne pathogens must begin before produce is even planted, by avoiding fields that have been subjected to flooding, on which animals have recently grazed or have been contaminated with manure. After planting, only clean potable water should be used for irrigation and harvesting equipment should be thoroughly cleaned and sanitized. Both field workers and packing-house and processing-plant personnel should be instructed in proper personal hygiene and provided with adequate sanitary and hand-washing facilities. Vehicles transporting finished products should be sanitized, properly loaded to provide adequate air circulation and maintained at appropriate temperatures. Likewise, retail display cases must be kept clean and at appropriate chilled temperatures. Finally, consumers should be informed as to proper hygienic handling of produce.

Minimizing Quality Loss During Fresh-cut Processing

Fresh-cut processing of fruits and vegetables represents a paradox in food science. Unlike other processes applied to food products, the unit operations applied to fresh commodities during fresh-cut processing tend to decrease shelf-life, rather than enhance the stability of the product.

Minimizing mechanical damage and microbial contamination

The quality and status of equipment used for peeling and cutting is critical in fresh-cut processing operations. Use of the sharpest cutting tools will extend product shelf-life. Dull utensils have been proven to cause excessive cell damage and bruising leading to poor quality. Severe peeling and cutting must be avoided.

Cantwell studied the impact of blade sharpness on the quality of fresh-cut melons stored at 5°C. The results indicated that melon chunks cut with a dull blade were susceptible to a translucence disorder, increased leakage and high ethanol concentrations in the package. Pear slices cut with a freshly sharpened knife retained their visual quality longer than fruit cut with a dull hand slicer. O'Beirne showed that slicing with a blunt blade enhanced the penetration of fresh-cut carrots by *E. coli* and its subsequent survival during storage. Frequent sharpening of machine and hand knives and proper cleaning and sanitizing of processing equipment and surfaces that come in contact with fresh-cuts is clearly a key control point in fresh-cut fruit processing.

Minimizing Transfer of Contamination during Washing Operations

Many large fresh-cut processing operations treat and recycle water in order to conserve this precious commodity. Care must be taken in recycling water so as not to introduce new risks of increased microorganisms to produce during washing. It is recommended that the best quality water be used for the final rinse of intact fruits and vegetables prior to fresh-cut processing.

Many operations inject chlorine as a disinfectant along with acid in order to maintain a pH range of 4.5-5.5 and assure the effectiveness of chlorine during washing operations. Measurement and recording of the chlorine level and the pH of wash water is a critical element of any quality assurance programme. When used to reduce the temperature of wash water, ice should be routinely tested in order to ensure that it is not a source of contamination.

The disinfectant level in wash water can be monitored through the measurement of oxidation-reduction potential (ORP). ORP is a measure of the oxidation level in the water in millivolts, and gives an indication of the efficacy of a sanitizer during processing. The stronger the oxidation, the faster the microbes are killed. Variables that affect antimicrobial activity during processing directly affect the ORP value and may also be used to determine the effectiveness of oxidizers such as hypochlorous acid, hypobromous acid, chlorine dioxide, ozone and peroxides.

Water quality can be maintained by:

— Closely following mixing directions for sanitizers;
— Using test strips or instruments to routinely measure sanitizer levels and pH;
— Removing debris; and
— Filtering wash water before recycling.

Temperature Management

Temperature control is essential at each step of a fresh-cut fruit process, in the distribution chain and in retail. Low temperature storage helps to slow the respiration rate, maintain quality and prolong shelf-life by keeping product temperature at the point where metabolic activity and microbial deterioration are minimized. Low temperature storage slows the growth of mould and bacteria. Thus the ideal storage temperature for each commodity should be researched carefully. Temperature is the most important factor in the preservation of fresh-cut fruit quality. The rates of chemical and biochemical reactions that affect quality are largely determined by temperature.

Heat and low temperature are both used in the pre-treatment of whole fruits and vegetables prior to fresh-cut processing in order to increase shelf-life. Mild heat treatment of whole apples before processing resulted in the retention of firmness during storage in some fresh-cut apple cultivars. Low temperature pretreatments such as the hydrocooling of asparagus spears and baby corn reduced respiration rates, reduced toughening of the texture and prolonged the quality of fresh-cut products.

Heated wash water treatments were shown to reduce pathogens on mango fruits. Mangoes treated with 100 ppm chlorine at 50oC had virtually no retention of aerobic bacteria on fruit skin and fresh-cut samples.

Many fruits of tropical or subtropical origin are sensitive to low temperatures. These fruits are injured after a period of exposure to chilling temperatures below 10 to 15°C, but above their freezing points. At these temperatures, fruit tissues are weakened owing to their inability to carry out normal metabolic processes. Various physical and biochemical alterations and cell

malfunctions occur in response to chilling stress. When chilling stress is continued, these alterations and malfunctions lead to the development of a variety of chilling injury symptoms, such as surface lesions, internal discoloration, water-soaking of the tissue and failure to ripen normally.

A number of fresh-cut fruit products, exhibit less sensitivity to chilling injury (CI) than intact fruits. The reasons why peeled tissue tolerates lower temperatures are unclear, but the degree of ripeness may play a role. Ripe fruit has been known to tolerate low temperatures better that unripe fruit.

Cantwell drew the following conclusions regarding the handling of fresh-cut products that are sensitive to chilling injury:

— It is important that intact chill-sensitive produce is not stored below the recommended storage temperature, before being prepared as a fresh-cut product.

— Once a chill-sensitive commodity is prepared as a fresh-cut product, storage at low temperature is needed to retard microbial growth and ensure quality.

— Microbial changes take place more rapidly than the appearance of any symptom of chilling injury.

— For chill-sensitive commodities, temperature and controlled atmosphere suitable for the intact produce are often not suitable for fresh-cut pieces.

Post-cutting Treatments

A number of physical and chemical treatments designed to delay physical decay processes in the tissues and to extend the shelf-life of fresh-cut fruits and vegetables have been developed and tested. Chemical methods rely on the inhibition of specific reactions that generate undesirable changes. Physical methods include reductions in temperature and/or oxygen concentration, the use of modified atmosphere (MA) and the application of heat or high pressures.

Chemical post-cutting treatments

Wounding and ethylene can initiate phenolic metabolism that ultimately leads to browning in fresh-cut tissue. Control of discoloration (pinking, reddening or blackening) or browning at cut surfaces is, therefore, a critical issue for fresh-cut producers. Outlined below are a number of strategies that may be used to reduce cut-surface discoloration and maintain the textural integrity of fresh-cut produce.

Acidification: PPO most effectively catalyses cut-surface discoloration at a neutral pH (around pH 7.0). Browning can, therefore, be slowed by dipping products in mildly acidic food grade solutions of acetic, ascorbic, citric, tartaric, fumaric or phosphoric acid. Quite often, combinations of acids (for example combinations of ascorbic and citric acid) are more effective than the use of acids individually. However, these acids may impart off-flavours or promote tissue softening, and must, therefore, be used with care.

Reducing agents: Reducing agents such as ascorbic acid or the erythrobate isomer of ascorbic acid are commonly used in the food industry to prevent PPO-mediated cut-surface discoloration. Ascorbic acid and erythrobate reduce PPO-induced discoloration at cut-surfaces by converting quinones (formed by PPO from phenolics) back into phenolic compounds. Once all of the ascorbic acid or erythrobate is exhausted, PPO browning will proceed uninhibited. Ascorbic acid or erythrobate are commonly used in solution at a concentration of 1 μM. Given their acidic nature, they may also reduce the surface pH of commodities, further slowing the rate of browning.

Application of edible coatings: The application of edible coatings such as sodium caseinate or stearic acid may be helpful in reducing white blush in vegetables such as carrots. Treatments that modify the water-retaining capacity of the cut surfaces also prevent white blush development.

Natural antimicrobials: Some plant extracts, such as ginger, marigold and cinnamon, exhibit antimicrobial activity. Cinnamon extract at 500 ppm was shown to control banana crown rot. Chitosan, a polysaccharide extracted from chitin, a natural substance in the shells of shrimp and crab has been shown to be effective in controlling rambutan fruit rot (*Lasiodiplodia theobromae*), mango anthracnose, banana crown rot and *Botrytis cineria* in strawberries. The application of vanillin as a dip at a concentration of 80 mN and storage at 5°C were effective in reducing the microbial populations in fresh-cut mangoes.

Firming agents: The firmness of the flesh of fresh-cut fruit products can be maintained through the application or treatment with calcium compounds. Dipping fresh-cut products in solutions of 0.5 to 1.0 percent calcium chloride was shown to be very effective in maintaining product firmness. Calcium ions form cross links within pectin chains, resulting in stronger cell walls. Calcium chloride treatments can, however, result in bitter off-flavours in some products.

Physical post-cutting treatments

Reduced oxygen: As PPO requires oxygen in order to induce cut-surface discoloration, reducing the oxygen levels in a fresh-cut product package by vacuum, modified atmosphere packaging (MAP) or gas flushing may reduce cut surface discoloration; however it does not completely stop it. Careful design of fresh-cut packaging is essential to assure the levels of oxygen within the package. While excessive levels of oxygen in a package may lead to discoloration of cut surfaces, inadequate levels may cause anaerobic metabolism, resulting in the production of off-flavours and odours.

MAP: Technologies continue to be developed in order to enhance the shelf-life of fresh-cut fruits and vegetables. Modified atmosphere and low temperature storage have been used for many years to reduce respiration rates and the deterioration of fresh-cut produce. MAs by definition are atmospheres that differ in composition from that of normal air. MAs can be generated passively or actively. In many developing countries, fresh-cut produce is generally packed in film bags or plastic containers overwrapped with film. A passive modified atmosphere is created within the package by the respiration process of the fresh-cut produce, combined with the permeability of the package. An active MA can be achieved by flushing with a particular gas or mixture of gases to create an initial atmosphere or by incorporating additives such as oxygen scavengers into the packaging.

Typically, MAs consisting of low oxygen and high CO_2 concentrations are used to reduce respiration rates and ethylene production in fresh-cut products. Such conditions can also be applied to prevent browning of the cut surface, retard enzyme activity and to inhibit microbial growth. A successful MAP design requires the determination of respiration rate and respiration quotient. The choice of appropriate polymer films with suitable permeability and appropriate film area; together with headspace volumes and knowledge of produce respiration characteristics, is essential. Excessive depletion of oxygen in the package can result in anaerobic respiration that leads to the formation of off-flavours.

Given the high respiration rate of fresh produce, hazardous anaerobic conditions and undesirable fermentation reactions can occur within packages of fresh-cut produce when stored under conditions of high temperature. During the retail of modified atmosphere packages, temperature fluctuations are unavoidable. The application of high oxygen concentrations (70 percent) could, however, overcome the disadvantages of low oxygen modified atmosphere packaging, for some ready-to-eat vegetables. High levels of oxygen are particularly effective in inhibiting enzymatic discoloration, preventing anaerobic fermentation reactions and in inhibiting microbial growth. It has been hypothesized that high oxygen levels may cause substrate inhibition of PPO, or alternatively, that high levels of colourless quinones subsequently formed may cause feedback inhibition of PPO.

Heat treatment: Heat treatments have been noted to improve tissue firmness in fresh-cut fruits such as apples slices and mango cubes found that heating whole apples at 45°C every 30 minutes for a two-hour period before slicing improved the texture of apple slices as compared to untreated apples, but this improvement depended upon the cultivar.

A 90-second heat-shock at 104°F (45°C) prevented wound-induced browning of iceberg lettuce. A low heat treatment to activate the enzyme pectinesterase prior to a $CaCl_2$ dip was helpful in preserving fresh-cut fruit texture. Pectinesterase is known to de-esterify pectin, thereby increasing the number of calcium binding sites.

Practical Aspects of Fresh-cut Processing

The Fresh-cut Chain: Harvest to Market

When preparing fresh-cut fruits and vegetables and root crops, the key steps of the chain: harvesting, receiving, pre-cooling, washing and disinfecting, peeling and trimming, cutting into specific sizes, sorting for defects, washing and cooling, packaging/labelling and storage and distribution.

Harvesting

Individuals engaged in fresh produce harvesting must be trained in the selection of fresh produce of sound quality. Produce must be handled in such a way as to avoid damage and contamination. Handlers must ensure that only the best quality produce is selected for fresh-cut processing. Management should supply all the equipment needed to minimize damage of fruits and vegetables during harvest, for example sharp harvesting knives, bags and padded boxes. Workers must wear clean clothes and observe sanitary measures. Mechanical harvesters are used in large operations to take handling closer to the source in the field, improving shelf-life and maintaining produce quality.

Receiving at the processing facility

The receiving point for fresh fruits and vegetables in a fresh-cut processing plant is a key quality check-point that should be monitored closely. Employees receiving produce should check for defects and make note of any problems that can be relayed to the grower. At this point a decision can be made on whether or not to process the produce. Large processing operations have a list of quality parameters and measurements that are recorded when fresh produce arrives at the processing facility. An inventory of stocks begins with receipt of the produce. Large processing facilities usually have an enclosed, refrigerated dock facility for receiving produce.

Recording as much information about the produce as possible would help to trace it back to the field in the event of a food crisis. Thus field number, harvest crew, vehicle registration and other information can be recorded when produce is received at the back door of a fresh-cut processing plant.

Pre-cooling

Ideally, produce should be cooled to remove field heat prior to storage or processing. Cooling extends the shelf-life of the final fresh-cut product. Large fresh-cut processing operations make use of any of a number of technologies such as chilled water baths, forced air cooling, vacuum cooling and packing with ice-water mixtures, for removing field heat from fresh produce.

While small operations may not be able to afford expensive cooling equipment, storing produce in a refrigerated area is adequate for cooling small quantities of produce. Care should be taken to note temperatures below which chilling injury of various fruits and vegetables would take place. While this research has been conducted for many temperate produce items, it is still not widely known for some tropical fruit and vegetables.

Intact fresh produce should be segregated in a small cold storage area to avoid cross contamination between it and that which has been pre-cut and washed. Staff should use a First-in, First-out (FIFO) policy for managing stock rotation.

Washing and disinfection

It is important to wash produce as soon as possible after harvest to remove damaged tissues. Water flumes and tanks are used in large operations to wash fresh produce prior to cutting and trimming. Potable water is a key requirement for washing in order to preclude the transfer of contamination from water to the produce.

Many large fresh-cut processing operations treat and recycle water to conserve this precious commodity. Care must be taken in recycling water so as not to present a new risk of increased microorganisms. Recycled water must be continuously treated and monitored. In situations where the organic material in wash water increases, the antimicrobial agent HOCl loses its effectiveness in maintaining water quality.

In very small operations, washing in a sink or under a tap with running water is appropriate. Produce may also be washed at a small scale in large shallow tanks that allow operators to move the produce freely through the water. In situations where concrete tanks are used they should ideally be tiled. Frequent changes of wash water are needed in order to effectively remove soil and other foreign matter from the produce. A better method is to continuously circulate wash water through a filter. Ideally several tanks should be used for washing operations. The first tank should be used for the removal of heavy soiling and subsequent cleaner tanks with chlorinated water should be used for final washing.

Large fresh-cut processing plants make use of continuous washers. In such situations, a moving conveyor picks up the produce and carries it under powerful sprays of water. Recirculation through a filter reduces overall water consumption. The temperature, contact time, pH and chlorine concentration should be monitored for their effectiveness during washing. The measurement of ORP can also be used to monitor the disinfectant level in wash water.

For vegetable wash water the following parameters may be used: temperature (0-5°C), pH (4.5-5.5), chlorine concentration (50-100 ppm), ORP (650-750 ppm). During the washing process any defective produce must be removed and discarded.

Peeling, trimming and deseeding

In small processing plants, knives are used to trim and peel fresh produce. Large processing plants make use of abrasive peelers and automated trimmers to accelerate the process. Available funding would determine whether a small processor could use technically advanced peelers and trimmers during operations. Automated peelers with abrasive rollers are used in some large processing plants for peeling potatoes and carrots. Machinery has been designed for specific high volume crops, including the use of high pressure air and lye and steam for peeling. However, for small operations, hand peeling is adequate.

Cutting operations

Produce may either be chopped, sliced, shredded, peeled, diced or sectioned. These operations are done mainly by hand in many small-scale operations. The dimensions of the finished product are determined by the end use identified for each produce item. Chopping boards and knives are used to create the desired size and shape for finished products. Research has shown that using a sharp

knife reduces physical damage to cut fruits and vegetables in that less stress is observed in the cells of produce cut with a sharp knife. Staff should ensure all knives and chopping boards are cleaned and sanitized before use so as to minimize the potential for contamination. Good Hygienic Practice (GHP) must be strictly observed during cutting operations.

Employees involved with cutting operations should be properly attired with protective clothing, including gloves, aprons and hair nets. They should be well trained in the preparation of products to minimize damage. Products should receive minimal handling to avoid bruising and contamination due to excessive handling. Whatever the method of cutting, the sizing of the produce must be uniform; given that non-uniformly sized products are not appealing to consumers.

Sorting for defects

Defective products may also hasten spoilage and decrease shelf-life. Removal of defects improves uniformity of the finished fresh-cut product and enhances shelf-life.

Rinsing of pre-cut fruits and vegetables

It is recommended that only water of the highest quality be used for the final rinse of pre-cut fruits and vegetables. Many operations inject chlorine as a disinfectant along with acid in order to maintain a pH range of 4.5-5.5 and to assure the effectiveness of chlorine. Measurement and recording of the chlorine level and the pH of wash water is therefore a critical element of any quality assurance programme. Ice used for reducing the temperature of water should be routinely tested in order to ensure that it is not a source of contamination.

Dipping

Produce can be optionally dipped in a solution of an acidulant/antioxidant blend consisting of a combination of ascorbic acid/citric acid for example, or in an anti-softening agent such as calcium chloride.

Drying

Excess water or liquid associated with the produce must be removed prior to packaging of fresh-cut products. Water in the finished product encourages mould growth and the growth of other microorganisms resulting in rapid deterioration of texture. Various manual and mechanical methods have been developed for the removal of excess water from fresh-cut fruits and vegetables. These include:

— Use of conveyor shakers to remove water through a mesh. The cut fruit or vegetable vibrates on the conveyor mesh.

— Air drying on conveyors with forced air or polar wind used to blow excess water off the surface of the wet produce.

— Spin-drying baskets, both automated and manual, that make use of centrifugation to remove excess water.

Packaging/labeling

Packaging facilitates the delivery of fresh-cut products of good quality to the consumer. Packaging protects products from physical damage and prevents physical and microbiological contamination. Packaging can on some occasions, as in the case of MA, delay spoilage of products. Cut fruits and vegetables deteriorate more rapidly than intact or packaged products.

Attention must be paid to the compatibility of produce, where mixed fruit and vegetable combinations are packaged. Ethylene-sensitive and ethylene-producing fruits should not be mixed, as this could result in water soaking and rapid deterioration of ethylene-sensitive produce.

Packaging formats for fresh-cut fruits and vegetables include plastic bags, thermoformed containers with film overwraps and rigid plastic containers. Packs are filled by hand in small operations, while automated equipment is used in larger operations. A metal detector is generally positioned at the end of the line in automated operations and bags of products pass through the detector as a control measure to test for metal contaminants. Finished product containers are labelled with a *'Use By'* date to alert customers of the optimum product shelf-life. Other company codes may be printed on the package.

MAP reduces the respiration rate of produce and thus slows the rate of spoilage. In addition, it creates anaerobic conditions or high carbon dioxide and low oxygen levels in the pack, to extend fresh produce shelf-life. Oriented polypropylene (OPP) is generally used in the MAP of fresh-cuts. Other packaging films used include perforated, thin, low density polyethylene (LDPE), monolayer polyvinylchloride (PVC) and ethylene vinyl acetate.

Managing and measuring quality during fresh-cut operations

Consumer acceptance of fresh-cut produce is influenced by colour, texture, aroma and taste. Convenience is a major reason for the procurement of fresh produce by consumers. Consumers choose fresh-cuts based on freshness, price and packaging. Their eating experience (aroma, taste, texture) determines whether consumers will continue to purchase such products.

A well-managed quality assurance programme should be developed to ensure that only the finest quality product leaves the fresh-cut processing plant. Management must develop specifications for the product at every stage of the fresh-cut processing chain: raw materials, in-process products and finished products. Monitoring programmes for storage and transport (temperature, relative humidity etc.) should also be developed. When raw material enters the plant, its quality determines the quality of the final product.

Quality assurance systems

Product quality is a prime criterion for gaining access to competitive markets. Quality control ensures that raw materials and finished products are handled, stored, processed or packaged to the required quality standards. The fundamental purpose of a quality assurance programme is to have timely and dependable information on all the attributes of a product which affect its quality. The basic functions of a quality assurance programme include:

— Physical, chemical and sensory evaluation of raw materials and fresh-cut products;
— In-process control of raw materials and fresh-cut products:
  • raw materials, ingredients and packaging supplies
  • processing parameters
  • finished products
— Microbiological analysis and control of raw materials and finished products;
— Control of storage and handling conditions;
— Control of sanitation and waste products;
— Assurance that final products are within the established legal and marketing standards.

Quality as a synonym for food safety can be used as a tool for marketing fresh-cut products in countries with high food safety standards. Product quality can be maintained throughout the supply chain by applying good practice. Methods for measuring each quality parameter should be clearly outlined, for example instrumentation and testing procedure. Staff should be assigned to monitor quality and to record the results of quality checks in daily log sheets. Management should verify daily log sheets to ensure that only products that meet specifications are packaged.

Raw material quality assessments might include:

— Measurement of the sugar content (degrees Brix);
— Measurement of the pH and organic acid content;
— Pressure testing to assess softness;
— Measurement of the colour at harvest;
— Observation for defects; and
— Verification that the product's shape is ideal for processing.

In-process quality measurements include measurements to ensure:

— Size of fresh-cut pieces are within specification;
— Products are mixed in correct proportions, if applicable;
— Products are free from excess water - measure the spin-drying cycle; and
— Product temperature is cool enough to promote product quality - measure and record temperature in the process.

Finished product quality measurements include assessment of:

— Packaging integrity;
— Average weight of packages;
— Label printing;
— Best Before/Use by dates and production dates are correct on the label; and
— The package temperature is adequate to sustain quality.

Managing and Maintaining Quality during Marketing

Fresh-cut fruit produce that is not produced on site is either transported on ice or in refrigerated trucks. Air transportation is used where markets are distant, primarily overseas markets. The fresh-cut products are normally transported in closed containers or netted/stretched film-wrapped pallet loads. The potential for temperature abuse and losses in quality are high during marketing and are a critical consideration.

Quality loss is a function of both time and temperature. If products are neglected, even for short periods at high temperatures during loading and off-loading, the shelf-life is reduced. Refrigeration removes excess heat and facilitates temperature control for fresh-cut produce during storage and transportation. Maintaining the 'Cold Chain' is the key to delivering wholesome fresh-cut products to the end user.

Fresh-cut fruits must be stored at the lowest temperature possible to maintain their freshness and minimize losses. In view of the fact that fresh-cut products are under severe physical stress during

processing, they should be held at a lower temperature than that recommended for uncut commodities. The optimal storage temperature for fresh-cut produce is, in general, set as the minimum required to minimize microbial growth while preventing chilling injury.

Relative humidity (% RH) is an indication of the balance between the water evaporated from the stored produce and its removal from the air by evaporating fans of the refrigeration system. RH levels of 80 to 90 percent are generally recommended to prevent a build-up of condensation that attracts mould and mildew in the cold room. Factors to consider when storing produce are the quantity of product to be stored; the type and method of packaging; stacking patterns to promote air circulation, air motion; system running time; and the power of the refrigeration unit. A still cold storage would take a longer time to cool products than one with fans to keep air circulating.

Assuring Safety in the Fresh-cut Processing Chain

Fresh-cut produce is highly vulnerable to contamination during processing. Every effort must, therefore, be made to minimize the risk of contamination in the producer-consumer chain. Factors in minimizing risk in the production-consumption chain are summarized below:

Harvesting and post-harvest handling

— Cleaning and sanitizing harvesting equipment;

— Excluding wild birds and animals from packing houses;

— Avoiding cross contamination during delivery to the processor; and

— Minimizing cutting and bruising.

Fresh-cut processing

— A programme for sanitizing surfaces and machines;

— Good preliminary decontamination and inspection;

— Avoiding severe peeling/cutting;

— Eliminating/minimizing human contact with processed products;

— Effective washing/anti-microbial dipping; and

— Avoiding post-dipping contamination.

Packaging/distribution/retail

— Careful selection of packaging material;

— Monitoring microbial quality of packaged products;

— Ensuring that the temperature is $<4^{\circ}C$;

— Modest shelf-life labeling;

— Suitably-designed vehicles
— Proper vehicle loading practices;
— Chill cabinet loading; and
— Education of retailer and consumer.

HACCP for Effective Running of a Fresh-cut Plant

From 1988 to the present day, HACCP principles have been promoted and incorporated into food safety legislation in many countries around the world. The approach was originally derived from Engineering System's – 'Failure Mode and Effect Analysis'. It was further developed by Pillsbury/ NASA for the American Space Programme. HACCP reduced the risk of astronauts suffering from the effects of consuming contaminated food whilst in space. It is a food safety methodology that relies on the identification of Critical Control Points (CCPs) in food production and preparation processes. Closely monitored CCPs will ensure that food is safe for human consumption.

HACCP is used to avoid the antiquated approach of testing the finished product without knowing the risks involved in preparation.

The HACCP system consists of seven principles:

- *Principle 1:* Conduct a hazard analysis. Three main categories of hazards: biological, chemical and physical, may affect products. Management should determine the food safety hazards and identify the preventive measures that can be applied by the plant to control these hazards.
- *Principle 2:* Identify CCPs, the steps or procedure in a food process at which control can be applied. The result is prevention of a food safety hazard or reduction to an acceptable level.
- *Principle 3:* Establish critical limits for each CCP. A critical limit is the maximum or minimum value to which a physical, biological or chemical hazard must be controlled at a critical control point to prevent, eliminate or reduce to an acceptable level.
- *Principle 4:* Establish CCP monitoring requirements. Monitoring activities are necessary to ensure that the process is under control at each critical control point.
- *Principle 5:* Establish corrective actions. These are actions to be taken when monitoring indicates a deviation from an established critical limit.

The final rule requires a plant's HACCP plan to identify the corrective actions to be taken if a critical limit is not met. Corrective actions are intended to ensure that no product injurious to health or otherwise adulterated as a result of the deviation, enters commerce.

- *Principle 6:* Establish record-keeping procedures – a written HACCP plan and records documenting the monitoring of critical control points, critical limits, verification activities and the handling of processing deviations.
- *Principle 7:* Establish procedures for ensuring that the HACCP system is working as intended. *Validation* ensures that the plans do what they were designed to do; that is, they are successful in ensuring the production of safe products. Plants will be required to validate their own HACCP plans. *Verification* ensures that the HACCP plan is adequate and is working as intended. Verification procedures may include such activities as a review of HACCP plans, CCP records, critical limits and microbial sampling and analysis.

In developing an HACCP plan, the whole process must be evaluated for possible risks. A HACCP flow chart would show the CCPs and a decision-making tree would be used to determine which hazard may actually be controlled and measured. CCPs of final rinse water and metal detection can be monitored and controlled if they are out of specification. This is the requirement for a hazard to qualify as a CCP.

Equipment Requirements for Fresh-cut Processing

Fresh-cut processing can be performed at different scales of operation. Regardless of the scale of operation, individuals engaged in fresh-cut processing should be properly attired with protective clothing, including gloves, aprons and hair nets. They should be well trained in the preparation of products to minimize damage, excessive handling and contamination.

Equipment must be cleanable, properly maintained and sanitized prior to use. Surfaces that come in contact with the food must be non-toxic, non-reactive with the produce, non-contaminating to the produce, non-corrosive and cleanable. Stainless steel of the 300 series is preferred for food contact surfaces.

Equipment for Fresh-cut Processing

Equipment for fresh-cut processing should be designed and constructed such that it is easy to clean and maintain, thus minimizing the potential for microbial contamination of the fresh-cut product.

The USFDA recommends the use of smooth, non-absorbent, sealed and easily cleanable food contact surfaces that are sloped to drain freely and made of durable, non-corrosive, non-toxic materials for fresh-cut processing. Food contact surfaces include items such as knives, conveyors, belts, chutes, product totes, gloves, tools including shovels and racks, cutting boards, tables, driers and spinner baskets as well as packing scales.

All food contact surfaces should be smoothly bonded (i.e. free of pits, folds, cracks, crevices, open seams, cotter pins, exposed threads and hinges) to avoid harbouring pathogens. Where two food contact surfaces meet, a cover over the juncture should be used to prevent food debris from collecting in the crevice and creating an area that is difficult to clean.

Equipment for small-scale manual operations

Requirements for a small processing plant include sharpened knives, plastic cutting boards, stainless steel or plastic bowls and tubs for storing trimmed products prior to cutting or packaging. These simple kitchen utensils are easy to obtain.

Many fresh-cut processing plants for leafy vegetables use automated equipment for all stages of fresh-cut processing. These include the following items of equipment:

— Product bin dumpers;
— Elevator belt conveyors;
— Automatic sorting machines;
— Abrasive peelers or high pressure water peelers;
— Trans-slicers or other cutting machines;
— Continuous washing conveyors;
— Spin driers, air driers, shakers;
— Form-fill-seal packaging machines;
— Weighing, labelling, coding machines;
— Boxing and palletizing machines;
— Metal detectors.

Maintenance of Equipment

Establishment of a preventive maintenance programme helps to ensure proper functioning of all equipment. Failure of equipment during production may increase the risk of microbial contamination, particularly from *L monocytogenes.*

Preventive maintenance includes the periodic examination and maintenance of equipment such as valves, gaskets, O-rings, pumps, screens, filters and heat exchanger plates. Appropriate action plans should be developed by small processers in the event of malfunctioning of important equipment, such as refrigeration equipment, disinfectant delivery systems, power systems or alarm systems. The following practices are also recommended:

— Maintenance and calibration of equipment by appropriately trained personnel. Maintenance personnel who work in the processing or packaging areas should be knowledgeable of, and comply with the hygiene requirements for production staff.

— Frequent sharpening of knives, if used, including retractable knives, and disinfecting them prior to use. Knives should be replaced if damaged or if they cannot otherwise be maintained in a sanitary condition.

— Frequently inspecting cutting blades and belts during processing operations for damage, product residue build-up or cleaning needs. Blades should be removed and separately cleaned. Remaining equipment parts must be disassembled and cleaned on a regular basis.

— Installation, calibration and maintenance of temperature measuring or recording devices to ensure their accuracy.

— Operating of metal detectors in accordance with manufacturers' instructions and checking for proper functioning at least on a daily basis to ensure effective detection of metals and removal of affected products. Procedures should be in place, such as the use of metal detectors during packaging operations, to minimize the possibility of metal entering finished product packages.

— Calibration of safety control devices that are essential for maintaining the proper level and activity of wash water disinfectant, at a frequency recommended by the manufacturer and documentation of this activity on instrument calibration forms/logs.

— Examination of air filters for both intake air and compressed air and changing at least as often as the manufacturer specifies, or more frequently if a problem is indicated, such as evidence of filter fouling or perforation.

Traceability of Fresh-cut Products

Traceability in Fresh-cut Chains

According to the Codex Alimentarius, *traceability* or product tracing is "the ability to follow the movement of a food through specified stages of production, processing and distribution". This definition encompasses two concepts: tracking, which refers to the ability to determine in real time the exact location and status of produce in the logistics chain; and tracing, which refers to the ability to reconstruct the historical flow of produce based on records maintained through the chain. The objectives of a good traceability system should be to improve supply management and aid tracing of food safety and quality parameters.

The benefits of a well-run traceability programme include:

— Improved efficiency at all levels within a company thus resulting in revenue generation.

— The ease with which the source of a safety or quality problem can be isolated and managed.

— Reduced risk of bad publicity, lawsuits and recalls.

— Reduced risk of rejection because unsafe products are not shipped and are quarantined thus contributing to the speed with which corrective actions is increased when food safety and quality problems are identified.

— Information retrieval is facilitated during a product recall or food-borne illness outbreak.

Many large restaurant chains and retailers now expect their suppliers to create traceability systems. These systems are verified through third party audits of their overall food safety system. A traceability programme should record information from the field to the retail display. Stringent legislation, consumer concerns about food safety and growing pressure from retailers have forced food manufacturers to look at every possible means of ensuring traceability and efficiency throughout the supply chain.

Record Keeping

Minimal farm information to be recorded for a traceability system includes:

— Plot number or some form of identification for land where the crop is grown.

— Records of Good Agricultural Practice (GAP):
  - pesticide application
  - fertilizer application
  - irrigation and water testing
  - land history
  - neighbouring land use.

— Harvest crew records:
  - records of training
  - reports of illness.

— Dates of harvest:
  - special conditions during harvest.

— Farm equipment and trucks used in the production and transport of fresh fruits and

vegetables:
  - name of driver
  - vehicle registration number.

Minimal packing house or fresh-cut processing records to be recorded for a traceability system, include:

— Quality assurance records of the fresh commodity on arrival at the packing house or processing facility.

— Records on fresh-cut processing staff:
  - records of training
  - records of illness.

— Good manufacturing practice (GMP) records:
  - records on sanitation
  - records on pest control
  - records on equipment maintenance

  - records of water tests
  - records of temperature during processing.
- In-process monitoring records of quality assurance.
- Packaging identification:
  - supplier
  - lot number
  - date of intake.

Distribution information to be recorded for a traceability system includes:

- Vehicle identification:
  - vehicle type
  - plate number.
- Container:
  - temperature
  - cleanliness
  - non-conformances.
- Driver identification:
  - names and contact numbers of drivers.
- Shipping company:
  - name, address and location of company.

Monitoring Traceability

In the past, a traceability code would be written or printed on each box and depending on the clarity of information in the code, a company could be contacted to review information associated with it. Only limited information would be contained in a hand-written code, for example the date of harvest or perhaps the lot number.

Bar coding has improved the way product movement is managed. A unique bar code is prepared for products harvested on farm. This bar code stores information on the grower, farm lot number, day harvested and crop harvested. A bar code placed on a bin of fruit after harvest could then be scanned at the packing house at intake. This code stays with the bin until it is repacked into a retail package. If products were stored before repackaging, the warehouse management system would assist in stock rotation based on information received from the bar code. When products are repacked into consumer packages, a new bar code is placed on the product to assist in tracing finished products back to the field in case of an emergency.

The Uniform Code Council (UCC) takes a global leadership role in establishing and promoting multi-industry standards for product identification and related electronic communication. International standards in bar coding include the use of a Global location number or GLN. This is a number that identifies any legal, functional or physical location within a business or organizational entity. The GLN contains 13 digits, all of which must be processed. GLNs can be presented in bar code format and can be physically marked on to trade units in order to identify the parties (buyer or supplier) involved in the transaction.

Some sectors of the food industry have moved on to the new technology of radio frequency identification (RFID) systems. RFID is a technology that makes use of grain-sized computer chips to track items at a distance. Each tiny chip is connected to an antenna that picks up signals beamed at it from a reader device. On picking up a signal, the chip communicates its unique identification number to the reader device, allowing the item to be remotely identified. 'Spy chips' as they are sometimes referred to, can beam back information anywhere from a radius of a few inches to 30 feet (~10 metres) away.

Under the USFDA Recording and Reporting Rule of the Bioterrorism Act, 2002, food companies must provide the USFDA with source data, including immediate previous source (IPS) and immediate subsequent recipient (ISR) for every component of a food item. Many companies have developed software databases and information systems with the ability to store detailed crop history and distribution data. European Union legislation for traceability will drive the industry in the European Union even further to develop adequate information systems.

Challenges for the fresh-cut produce industry

Capturing a change in the lot number of produce as it moves on a production line may prove difficult. Furthermore, mixed lots of products in a single package (a fruit salad, for example), complicate traceability, particularly if components of the mix are sourced from different growers. In the event of a crisis, large quantities of products might have to be recalled if the situation is not properly controlled.

Small producers on a limited budget may not be able to afford bar coding systems. However, a detailed number coding system with records to support traceability could provide sufficient information in the event of a crisis.

Layout and Maintenance of a Fresh-cut Processing Facility

The fresh-cut processing plant

Choosing the site of a fresh-cut processing facility

The following issues must be considered when deciding on the location of a fresh-cut processing facility:

- Distance of the facility from the supply of raw materials;
- Accessibility of the processing facility;
- Accessibility to expected markets;
- Availability of local labour;
- Environment in which the plant is located - is the building situated in a clean area or is it in the vicinity of an area with much dust, waste or stagnant water?

Fresh fruits and vegetables are bulky and spoil rapidly. It is therefore best to locate the plant close to the area of production in order to reduce transport and handling costs. Excessive handling of fresh produce results in injury, leading to rapid spoilage. In view of the fact that markets for fresh-cut fruits and vegetables would most likely be located in urban areas, the plant should be located in the vicinity of a main road for easy access.

The fresh-cut processing facility should be located on cleared ground, away from sources of insects, rodents or foul odours. It should have a good supply of potable water, and if required, a reliable supply of electricity. A sealed road should be available to provide access for delivery of raw materials and packaging, and for shipping of fresh-cut products. This road should be well connected to the major road network. Good drainage of the site from front to back would be most desirable.

Sanitation design of a processing facility for fresh-cut produce

The external and internal design of the processing facility should facilitate cleaning and prevent product contamination. The surroundings of the facility should be planted with grass to serve as an efficient trap for air-borne dust. The internal walls of the building should be plastered or rendered with concrete. The surface finish of the walls should have no cracks or ledges, which could harbour dirt or insects. The lower walls of the interior of the plant should be tiled or painted with water-proof gloss paint to withstand splashing during equipment cleaning. The upper walls should be painted with good quality emulsion paint.

Natural daylight is less costly than electricity. As far as possible, efforts should be made to use natural light in the building. The size of the building and the number of windows would depend on the level of investment available for construction. Store rooms do not need windows, but require ventilation. All windows, vents and fans should be screened in order to prevent the entry of insects, birds and rodents. Buildings should be pest-proofed with no gaps under the doors. The floor should be curved to meet the walls in order to prevent the collection of dirt. Doors in the processing room should be kept closed unless they are fitted with thin metal chains or plastic curtains to prevent the entry of flies.

Operations must be kept sanitary by cleaning up spills as soon as they occur and by washing the plant thoroughly at the end of each day of production. Wire mesh should be placed over drains to prevent the exit or entry of pests. Washing and toilet facilities must be provided for employees, preferably in a separate building. If this is not possible, two closed doors are required between the toilet and the processing area, to prevent insects and odours from entering that area.

Window ledges should be sloped in order to prevent them from collecting dust, dirt or old clothes that may be left there by workers. All windows should be fitted with fly-proof mesh. Attention must be paid to other points through which insects, birds and rodents can enter the processing room or store-room. Gaps between the roof and the walls and gaps in the roof must be properly sealed. Power lines must be fitted with suitable discs at least 25 centimetres in diameter in order to preclude the entry of rats into buildings.

Floors must be designed so that they drain efficiently. One way to facilitate proper drainage is by sloping all floors toward a central drainage channel. The drain should be covered with a removable grating to facilitate cleaning. Drains are a favourite entry point for pests such as rats and cockroaches and thus drain outlets must be fitted with a removable mesh basket or something similar.

All electric power points should be fixed at least 1-1.5 metres high on the walls in order to ensure that they are not affected by washing operations; although expensive, waterproof power points are preferred in wet areas. Alternatively, they may be suspended above the machines they serve. All 3-phase equipment should be installed by a competent electrician. Good lighting should be provided for the working environment and any bulbs used should be protected from breakage with a shatterproof plastic diffuser or sleeve covers.

Good ventilation is essential and large mesh-covered windows, roof vents and ceiling fans may be used. A reliable supply of potable water for cleaning equipment and for washing operations must be ensured. In situations where water supplies are unreliable, strategies must be used to overcome water supply problems. A simple double-chambered high-level water tank can facilitate water purification by treating cloudy water with chlorine and allowing it to settle in one chamber of the tank, while the other chamber is being used.

The USFDA recommends that the processing facility and its structures (such as walls, ceilings, floors, windows, doors, vents and drains) be designed for easy cleaning and maintenance to protect the products from microbial, physical and chemical contamination.

Product contamination and cross-contamination within the plant can be minimized by designing the operation with sound construction materials and by managing the flow of material and people through the plant. Air flow through the plant must also be managed. Good sanitation should be monitored from receipt of products through pre-cooling, fresh-cut processing, packing and storage operations. Buildings, fixtures and equipment should be maintained in a clean condition that will protect fresh-cut produce from potential microbial, chemical and physical contamination.

The following additional recommendations for building design are provided by the USFDA:

— All exterior doors and entrances should be closed when not in use, ensuring an adequate seal when exterior doors and entrances are closed.

- Waste water should be promptly removed from the processing area and collected in a designated area for primary treatment to prevent product and equipment contamination.
- Overhead pipelines should be avoided as they could be a source of product contamination by overhead condensate. Catch pans may be used for overhead pipelines, but they should be cleaned on a regular basis.
- Construction materials made out of wood are not recommended for food contact surfaces as they harbour microbes. Wood splinters could also pose a physical hazard as they could contaminate fresh-cut products. Non-wooden construction material like metal or plastic are recommended because they also reduce the potential for microbial contamination.

Good Manufacturing Practices (GMPs)

GMPs are systems put in place to ensure that food prepared in a plant is sound and free of contamination. GMPs include:

- Food safety programmes;
- Management systems;
- Pest control programmes;
- Operational methods and personnel practices;
- Maintenance for food safety;
- Cleaning practices.

GMPs are monitored and audited by government agencies and private firms on behalf of retail clients. Each agency should have developed rigorous standards based on the main GMP principles. GMPs are a prerequisite to establishing an HACCP programme.

Cleaning practices for fresh-cut processing facilities

The overall cleanliness of a food-processing unit, however small, can have a major impact on the quality of finished products. For example, particles of food that are trapped in the corner of a tank or in a pipe, can allow the growth of microorganisms and this can cause major contamination of products during processing on the following day.

Management should develop the following:

— A master cleaning schedule for daily, weekly and deep cleaning activities.

— Detailed cleaning methods for each type of equipment and for each area in and around the fresh-cut processing area.

— Assignment and training of staff in cleaning methods and recording information.

A successful sanitation programme is one in which staff is well trained in cleaning methods and is assigned specific responsibility for each area of the processing facility.

Pest control for fresh-cut processing facilities

Pests such as rodents, birds and insects are known to harbour human pathogens. Pests can enter food packing and processing areas through cracks and crevices, thus rendering food susceptible to contamination. A well-planned, preventive pest control programme would help to deter pests from the food handling area.

Management is responsible for:

— Pest proofing of all buildings that house food - all cracks and crevices must be sealed, windows must be screened and the perimeter of the building must be kept clean and clear

— Conducting routine inspections for rodents and flying insects, and other pests, with the use of preventive measures like rodent bait, bird traps or mouse traps.

Large processing facilities generally employ pest control companies on contract to visit and conduct routine inspections of the facilities and to prepare a report at each visit. In small operations, one staff member could be trained in pest monitoring and control so as to facilitate the conduct of regular inspections and recording of information. Maintaining clean, sanitary surroundings helps in reducing the risk of pest entry.

Emerging and Growing Concerns of Fresh-cut Produce Industry

Consumers in developed countries are becoming increasingly concerned about the conditions under which their food is grown and transported. A growing need for ethical food production, traceability as well as preservation of the environment has been voiced by consumer groups to create public awareness.

Carbon footprints of fresh-cut produce

The cost of food production is lower in developing countries, where labour costs are a fraction of those in the developed world. Bananas are transported from Central and South America to Europe for consumption on a year-round basis. Other commodities are shipped or air-freighted over thousands of miles to provide food all year round. Developed countries are not subject to the limits of seasonality because fresh produce can be sourced from many areas to meet their needs for year-round supply.

'Food miles' refer to the distance food travels from the point of its production until it reaches the consumer or end-user. The food miles created by transporting commodities over long distances have raised alarm with some consumer groups and environmentalists, who believe that carbon emissions impact the environment.

Food travels an estimated 30 billion kilometres each year in the United Kingdom. This estimated distance includes travel of imports by boat and air and transport by lorries and cars. The transportation of food is responsible for the United Kingdom adding nearly 19 million tonnes of CO_2 to the atmosphere each year, contributing to a huge 'carbon footprint'. Over 2 million tonnes of this CO_2 is produced simply by cars travelling to and from shops.

One school of thought is that choosing food that is locally produced and seasonal would reduce the distance over which food must travel. Reducing the number of food miles could have a dramatic effect globally on the reduction of CO_2 emissions. However, in densely populated urban centres with little or no agriculture, the practice of local consumption would be very difficult to implement.

Waste management in fresh-cut processing

Another significant concern in developed countries is that of waste management. Many programmes are being developed to recycle and sort the enormous volumes of waste generated by consumers. The 'Green Movement', as it is called in some countries, educates customers on separating waste – glass bottles, paper, plastic, organic matter – so that waste may be recycled or broken down for use as manure. Cardboard or paper packaging is preferred over plastic packaging because it may be recycled and the board reused to form packaging. In fact plastic is equally recyclable, if not more so.

In addition there is the waste material produced from the processing of fresh-cut produce itself, the unwanted material that is peeled, trimmed and cut from the fresh fruits and vegetables during their preparation into fresh-cut products. Fresh-cut products must also be disposed of after reaching their use-by dates without being bought or consumed in time. All of these wastes can amount to a huge cost for a country.

In the United Kingdom alone, the volume of this wasted food amounts to about 6.7 million tonnes, which is disposed of in landfills each year. This food includes packaged foods that have never been opened. A new study says that reducing the waste of food in the country could prevent 18 million tonnes of CO2-equivalents from being emitted each year – the same as taking one in five cars off British roads.

This food waste might be converted into biofertilizer and renewable energy instead of sending it to landfills where it creates methane, a powerful greenhouse gas.

Factors that influence future growth of the tropical fresh-cut produce industry

Developed countries are leading the way in innovation in the fresh-cut produce industry. Government regulations, as well as large retail groups, are demanding higher standards of sanitary design for processing plants and packaged food. Automation and custom-made equipment are rapidly replacing hand-cutting operations.

Foreign bodies in products are being removed using automated detection systems. Chlorine, traditionally used as a disinfectant in wash water, is being replaced by a host of other chemical and non-chemical treatments as the by-products of chlorine are thought to be cancer-forming. Retailers demand a longer shelf-life and research is progressing in MAP and in ideal storage conditions to prolong shelf-life.

Future growth and development of the tropical fresh-cut industry will hinge upon:

— Principles and practice of fresh produce traceability;

— Microbiological issues related to fresh produce;
— Developments in the packaging of fresh-cut produce;
— Advances in sulphite-replacement technologies for fresh-cut produce;
— Safe and effective biocides for washing and decontamination; and
— Recovering value by reducing waste in the fresh-cut produce industry.

All of these concerns in the modern export market pose challenges which must be met by developing countries.

Processing of Fresh-cut Fruits and Vegetables in Thailand

Fresh-cut produce

Fresh-cut produce has been marketed by street vendors in Thailand for many years. Fresh-cut products are widely sold in wet markets and are increasingly sold in supermarkets across Thailand. A range of fresh-cut vegetables, including leafy vegetables (basil, unchoi, kale etc.), chilli, lemon grass, galangal, asparagus, baby corn, peanut sprouts and ready-to-cook vegetables is also exported.

The fresh produce sections of large Thai supermarkets contain a myriad of fresh and fresh-cut fruits and vegetables. A survey conducted by Sa-nguanpuag *et al.* reported that fresh-cut vegetables command a larger market share than fresh-cut fruits in regular and upscale supermarkets. This survey also determined that the price of fresh-cut produce was higher than that of intact fresh produce. Growth in the Thai fresh-cut industry has been largely due to increasing incomes and an increasing number of working couples and families who have little time to prepare their meals from scratch.

Processing of fresh-cut fruits

Fresh-cut fruits are produced at the small vendor level, at the supermarket level and by small processors in Thailand. Steps in the processes vary in accordance with the target market. Small vendors who target the mass market, generally retain produce on ice, while cutting on demand. Supermarkets on the other hand, which target the higher end consumer who is increasingly safety and quality conscious, cut and package fresh-cut fruits under hygienic conditions on a daily basis and display them on ice or under chilled conditions. Small processors who also target supermarkets and the food service sector, often include an anti-browning treatment as a processing step, given the lag time between production and marketing, and given the expected longer shelf-life of their products.

Processing of selected fresh-cut fruits for the export market

Fresh-cut papaya (Carica papaya)

Harvesting

Papaya should be harvested when approximately 75-80 percent of the skin of the fruit turns yellow. The fruit should be cut off the tree using scissors or a sharp knife. It is recommended that the stem be cut closer to the tree trunk in order to prevent damage to the fruit. The stem of the harvested fruit must subsequently be cut to approximately 1 inch. Fruits must be carefully packed in baskets and transported to the processing facility.

Grading and sorting

Papaya must be sorted for uniformity of size and color, firmness, freedom from defects such as skin abrasions pitting, insect injury and freedom from decay.

Fresh-cut pineapple (Ananas comosus L.)

Harvesting

Pineapples should be harvested in the early morning or late in the evening to protect fruit from the sun and to reduce the heat load on harvesting. Pineapples are harvested by snapping fruit from the stalk using a downward motion. The fruits should be placed in baskets upside down on the crown and while in the field, left under shade. The stems and the crowns should be trimmed.

Pineapples should be transported as soon as possible after harvest to the packinghouse and should be pre-cooled to remove field heat. This can be done either by submerging in water, or by air cooling. Should water be used, it should be chlorinated and frequently replaced. Fruits with a high level of flesh translucency can be separated at this stage as they sink on immersion in water.

Grading and sorting

Pineapples must be sorted for the required color stage of ripening and for uniformity of ripening. They must be fresh and clean in appearance, free of wounds from harvesting or handling, including scratches, punctures or bruises.

Fresh-cut pummelo (Citrus maxima or citrus grandis)

Harvesting

Pummelos should be harvested between 6.5 and 7.5 months after full bloom. Fruit harvested later than 7.5 months after full bloom are generally sweeter. Determination of the harvesting date is dependent on the sugar/acid ratio required by the customer and on the cultivar.

Pummelos should be harvested by cutting the stem of the fruit with scissors and dropping the fruit into a cloth bag. The harvested fruit is then transferred to a basket and kept in the shade, following which is transported to the processing facility and stored in a cool room at 10oC and 85% RH.

Grading and sorting

Fruits damaged by pests must be separated out.

Fresh-cut Watermelon (Citrullus lanatus Thunb.)

Harvesting

Watermelons should be harvested at 25-30 days after full bloom or when the ground spot (the portion of the melon resting on the soil) changes from pale white to a creamy yellow color. Experienced farmers or fruit sellers use the knocking of sound fruit as an indicator of the maturity index of watermelon fruits. Immature fruit give off a metallic ringing sound while mature fruit give a dull or hollow sound.

Fruit are harvested by cutting them off the vine using scissors. They are transferred to a basket and left under shade.

Grading and sorting

Watermelons must be sorted for maturity and for damage caused by bruising and cracking.

Processing of fresh-cut non-leafy vegetables

Fresh-cut vegetables are sold in wet markets and supermarkets and are packaged for export. Steps in the processes vary in accordance with the target market. Small vendors generally trim the produce prior to cutting and selling in the wet market under ambient conditions. Supermarkets on the other hand, generally wash the produce prior to trimming, cutting and packaging, under hygienic conditions. Vegetables destined for export are subjected to more stringent processing conditions, wherein they are trimmed, cleaned, cut, washed, sanitized, dried and packaged.

Processing of selected non-leafy vegetables for export from Thailand

Fresh-cut Asparagus (Asparagus officinalis L.)

Harvesting

Asparagus must be harvested according to length, diameter, and condition of tips and in accordance with customer specifications. In order to slow respiration, asparagus must be harvested either during the early morning or evening, when temperatures are cool.

Grading and sorting

Asparagus is graded into classes and also sorted for wholesomeness, freedom from damage caused by pests, freedom from any visible foreign matter, freedom from bruising, fresh in appearance and fresh-smelling.

Baby corn (Zea mays L.)

Harvesting

The harvest index for baby corn is indicated in various ways. Observation of the emergence of silk length could be one practical index. Baby corn must be hand-harvested 2 to 3 days after silking or when the silk emerges 1-2 centimetres from the top end of cobs, or when the plant is 45-50 days old. The top cob must be harvested first, followed by harvesting of the second cob after 1-2 days. Baby corn must be harvested over 7-10 days to avoid overgrowth of the cob and changes which lead to a fibrous texture.

Cobs must be properly cooled using forced-air cooling immediately after harvest. When possible, harvesting should be carried out in the morning, when the moisture content is highest and product and ambient temperatures are low.

De-husking

Baby corn cobs must be hand husked and the silk must be removed. In order to facilitate de-husking, a thin sharp knife should be used to open and peel off the husk. Care should be taken not to break or damage the cob in the process.

Grading and sorting

Baby corn must be sorted for wholesomeness, freedom from damage caused by pests, freedom from any visible foreign matter and bruising, fresh in appearance, and practically free of silk. The cut on the base of the cobs should be clean.

Fresh-cut mixed vegetables: baby pak choi, sweet pepper, asparagus and baby corn

Pak choi (Brassica chinensis)

Harvesting

Pak choi is harvested by hand at the age of 25-30 days in order to obtain young leaves. The head must be carefully cut at the base and deep in the soil, in order to leave the shoot system intact and undamaged and to reduce water loss. Harvesting should start from midmorning to midday. Following harvest, external and damaged leaves must be removed and care must be taken to avoid bruising during all operations. Pak choi must be transferred to a clean plastic basket or box with plastic over wrap and stored in a cool room prior to processing

Grading and Sorting

Pak choi heads are graded according to their size and the quality of their leaves.

Grade and sort for size and for intact leaves and stems

Clean cut the base of the head

Transfer pak choi to a basket and hydrocool

Dip the basket in chlorinated water (200 ppm) at 10ºC for 5 minutes

Free chlorine should be at a level of 80 ppm at the start of washing. At end of washing, the level of remaining free chlorine should be in excess of 20 ppm.

Use Forced Air Cooling

Sweet Pepper (Capsicum annuum L.)

Harvesting

Green peppers must be harvested when they are shiny and firm, usually at full color stage (light green color) for greenhouse-grown peppers and at the mature green stage (dark green color) for field grown peppers. Red peppers are harvested from the plant at a later stage when the fruit turns red.

Sweet pepper must be hand harvested using a thin, sharp knife to cut the fruit from the stem and to avoid bruising. Peppers should be cooled soon after harvest using either forced air cooling or hydrocooling. In situations where hydrocooling is applied, peppers should be immediately blow dried with air. Peppers are packed in a box and are transferred to a processing plant.

Grading and sorting

Sweet peppers are sorted for firmness, freshness in appearance and freedom from blemishes.

Fresh-cut Ready-To-Cook mixed vegetables: baby pak choi, sweet pepper, asparagus and baby corn

Ready-to-cook mixed fresh vegetables: (1) whole head of pak choi cut lengthwise between 8-10 cm, (2) green and red sweet pepper cut into strips (3 x 5 cm), (3) asparagus approximately 6 cm in length and (4) baby corn cobs with the husk, silk and stalk removed, between 5-7 cm and in width and 1-1.2 cm.

References

Bachmann, J. & Earles, R. 2000. *Post-harvest handling of fruits and vegetables.* ATTRA. 24 pp.

Beaulieu, J.C. & Gorny, J.L. 2001. *Fresh-cut Fruits.USDA Handbook.* 66: 16-30.

Kim, J.G. 2007. Fresh-cut market potential and challenges in Far-East Asia. Acta. Hort. 746, 53-60.

Siripanich, J. 2000. Post-harvest physiology of tropical fruits. *Acta Horticulturae,* 575.

Watada, A.E. & Qi, L. 1999. Quality of fresh-cut produce. *Post-harvest Biology and Technology,* 15(3): 201-205.

5

Fruit Juice Processing

Over the past several decades there has been a growing trend toward adding value to raw agricultural products. As populations have become more urban, this trend has accelerated. The need for stable, convenient foods has increased along with the demand for exotic products for international cuisine. Within the globalization of the food industry, the demand for quality juice and juice type beverages has markedly expanded. Traditionally, only a handful of fruit and vegetable juices have served this market as large multinational companies or their affiliates, have captured the majority of national and international juice trade. Juices such as orange, grape, pineapple, apple, tomato and blends are well established in developed countries. Now, minor juices, tropical juices and juice products are attracting new attention.

Trade and technical literature available reflect these trends. There is an impressive body of published plus increasingly, Internet information on all aspects of the juice and juice beverage industry. Juice processing technology ranges from individuals preparing juice at home for family consumption to multinational conglomerates with several interconnected high capacity plants and juice product lines serving global demands. However, there is far less attention paid to minor juices, small or local manufacture of such products and the specific problems faced by producers who have not shared in this growth.

There are some general principles that should be understood by anyone contemplating entering the juice processing business. Many aspects of production, postharvest handling, food safety, quality, unit operations, processing and packaging procedures, as well as regulatory control, are common to practically all juice products. Then there are a number of product-specific details, dependent upon the morphology, composition and character of the individual fruit. This publication will provide adequate details so interested parties can either improve existing operations or develop new processing establishments. Potential capabilities include adding value to local agricultural products, providing employment, augmenting the quality, safety, economy and diversity of the local food supply.

This entails a working understanding of food science and technology. In turn, it requires attention to the knowledge resources available though unevenly accessible globally. A good technical library with the latest texts, supplemented by complete journal holdings, abstract services and as their electronic equivalent, Internet linkages and on line search capability, is an expensive undertaking for even the most affluent institutions. Such facilities remain a remote possibility in developing countries. However equipment and access costs are dropping and the pace of information technology advances has never been faster and will have global implications.

Dramatic advances in information technology (IT) promise to greatly enhance technical information transfer in all fields, including Food Science and Technology (FST). When that time arrives, texts such as this will be on line with the key subject matter including hyperlinks to relevant references, sources of suppliers, national and international regulatory aspects and marketing information. Updating will involve periodically introducing new material and replacing obsolete information. The FAO Website is an excellent example of this evolving trend. However, hard copy texts provide the necessary insights and guideline for the near term.

Elegant technology does not negate the need for a fundamental understanding and working knowledge of the science and technology behind juice manufacture. Sophisticated processing machinery, computer data acquisition and process control will not replace hands-on experience with specific commodities. Close observation and manipulation of a crop from breeding advances and cultivation to final juice consumption under varying circumstances will always be necessary. It is made easier by technology, but it is not replaced by it. In fact, the experienced juice technologist with a comprehensive understanding of local raw material, global practices can improve the efficiency and profitability of a juice operation.

International visitors to modern juice processing facilities are often in awe of the elegant, sophisticated, costly operations they view, unaware or forgetting the tremendous investment in human capital required to arrive at that point. Guests from even the poorer regions of the world should recognize that they possess untapped human resources, potentially skilled labour and, most importantly, an agricultural environment where many foods can be grown that are impractical to grow or process in the temperate zone.

Visitors may then recall vast heaps of fruits in their own country going to waste due to lack of markets or processing outlets. The logical conclusion is to consider developing a local juice processing industry based on the impressive operation they are viewing. This is, still, putting the cart before the horse. The many logistical and infrastructure hurdles, which must be overcome before the first fruit is processed will not be evident to most visitors.

Indeed, the rusting hulks of once modern food processing plants can be found in many regions of the developing world, a testament to inadequate planning by the international assistance community and their local counterparts.Fruits are ubiquitous in most temperate and tropical zones. There may be ample raw materials available for short seasons, which mostly goes to waste due to the lack of processing facilities. There are opportunities for small and medium scale producers who are unlikely to compete directly with major international firms. Yet there are untapped national markets. With a realistic attitude, attention to competitive advantages and a strong business commitment, growth including export opportunities are achievable goals. Challenges are great, but success has been achieved in the past and will be again in the future. We hope readers of this text will be among the successful ones.

Historical Background

The manufacture of juices from fruits and vegetables is as old (or older) than agriculture. During the ripening process most fruits soften to the point where simply handling or transporting them yields more juice than flesh, albeit often partially fermented. The resulting pulpy fluid, easily separated from seed and skin, is generally more flavourful than the more solid portion. Hunter-gatherers could either consume quasi juices directly or collect the soft fruit in reasonably leak-free containers for later use. Simply transporting the fruit any distance soon guaranteed juice. Except for cool temperature climates, storage life was limited to mere hours before incipient fermentation modified the character of the juice appreciably.

Through trial and error humans learned practical ways of extracting juice from various sources and, most importantly, which attractive but toxic fruits to avoid. Tool making skills fostered the manufacture of devices for macerating fruits and extracting juices. Containers for storing foods, including receptacles for fluids were also devised from local resources, i.e. woven plant fibre and wood, clay soils and animal skin/intestine. Such methods and equipment can still be found in isolated pre-industrialized regions. Otherwise they are seen only in museums as prehistoric or early artifacts of extinct civilizations. Until fairly recently, liquid containers were quite primitive compared to those of the 20th century.

The perishable nature of juices dictated immediate consumption within less than 24 hours in warm climates and extended but still limited time in cooler environs. Natural chilling or freezing was the only alternative to microbial modification of the juice. Such fermentation was the basis for wine. It was found that the juice, after bubbling mysteriously, has a distinctly different character and affect on those consuming the product. Also, the material stored several days longer slowly turned harshly acidic or developed an even more unpalatable surface growth and taste. This early vinegar was used as a preservative for other fruits and vegetables. Grape domestication followed this discovery.

With the development of agriculture over the last 10 millennia, the cultivation of crops provided a fairly reliable source of food, including fruits appropriate for juice and beverage use. Unless the juice was consumed fresh, soon after pressing, fermentation was the likely, often desired consequence until preservation techniques were developed. In fact, the concept of maintaining a fruit destined for juicing in its whole, intact form until the juice is needed continues to be a sound principle. Even today, maintaining the fruit intact is one of the easiest ways of preserving juice quality.

Value of Juice

The global market for juice and juice products was estimated to be about 50 billion litres in the late 1990s. In the United States of America alone, the retail commercial value of the almost 20 billion litres of juice and juice products exceeded US$18 billion, roughly 3 percent of a total food sales expenditure of US$630 billion. World trade has accelerated over the last decade with developing countries achieving over 60 percent of fruit juice exports. Brazil, the largest citrus producer, accounts for about 25 percent of world production. Juice utilization in decreasing order is: Orange, apple, grape and pineapple.

Despite the more perishable nature of juices, there are many practical reasons for their manufacture, processing and increased consumption:

- The more delicate, soft fruits cannot be kept intact over long periods and tend to
- before or at harvesting. Juicing is the logical alternative,
- Even more durable fruit may have poor size, shape or blemished portions that preclude marketing them as fresh whole fruit. Upon trimming and careful inspection, the sound portion can be juiced. In this text, it will be continually emphasized that unsound fruit or those undergoing incipient spoilage are not recommended for juicing. However, use of sound culls represents a sensible, technically and economically valid utilization strategy,
- Juices can be consumed more conveniently than whole fruits. Envision a motorist in speeding traffic (cell phone in hand?) or a mother holding an active child, both motorists attempting to peel an orange vs. handling an easily opened juice container. "Dashboard dining" is a way of life for many, therefore juice products and clever packaging play a key role,
- The very young, elderly and infirm may have problems eating, let alone peeling certain fruits. Drinking juice is an effective, nutritious alternative,
- In addition to the nutritive value of juices and the additional health benefits from phytochemicals, recently recognized components of many fruit juices will
- the popularity of such products,
- Even flavourful juices, that may not be balanced nutritionally or which lack
- nutrients or phytochemicals, can be blended or act as effective carriers for other natural or synthetic nutrients such as vitamins, minerals and
- Fluid food products, including juices are easier to process, to heat, cool, freeze, standardize, transport, etc., than solid foods or fluids containing particulates. Thus processing efficiency, safety and quality norms are easier to meet,

- The ease of mixing fluids promotes development of unique blends of juices with other products and practical combinations not found naturally,
- Highly flavoured juices are the basis for a range of juice co-products, such as
- ice cream and confectionery flavours, smoothies, bakery ingredients, etc.,
- Modern processing, packaging, ingredient technology and distribution systems insure safe, stable and appealing juice and beverage products in a convenient, economical form far from the raw material source or season.

Perusal of food markets the world over, suggest that juices of all types and in all forms have an important role in both food nourishment and enjoyment. This trend has certainly accelerated over the last half decade.

Defining Juices

Juice is defined in the most general sense as the extractable fluid contents of cells or tissues. Although many fruit juices are the obvious result of expressing the liquid from the whole or cut fruit, there are some fruits where the distinction is not so apparent. For example, squeezing peeled mango flesh yields little juice, until the flesh is comminuted. Even then a thick puree is the end result. In contrast, comminuted apples yield a readily expressible juice. Adding water to mango puree would decrease consistency, but it is not considered juice. The fluid expressed from lemons, limes and excessively acid fruits is certainly juice, but the liquid is too sour to consume directly without dilution with sugar and water to produce lemonade or limeade.

Codex Alimentarius defines juice as "unfermented but fermentable juice, intended for direct consumption, obtained by the mechanical process from sound, ripe fruits, preserved exclusively by physical means. The juice may be turbid or clear. The juice may have been concentrated and later reconstituted with water suitable for the purpose of maintaining the essential composition and quality factors of the juice. The addition of sugars or acids can be permitted but must be endorsed in the individual standard.".

Tradition and manufacturing procedures dictate the distinction between juice, puree and pulp. Juices may be prepared from nearly all fruits, if desired, but purees or pulp may serve commercial needs more economically.

Banana is easily pureed, but more effort and expense is required to produce a clear juice from the pulp. Even a fruit as refractory as mamey zapote that yields very little free juice can be extracted with added water and then concentrated back to the starting fruit soluble solids level. Strategies for dealing with difficult or impossible-to-juice produce will be examined later.

It is absolutely necessary for someone starting a juice manufacturing operation to become familiar with the regulations and requirements of their market. For commercial purposes it is important to define the differences carefully and insure that specifications and labelling are correct. There are circumstances where a 100 percent juice or puree product is impractical while dilution with other juices and/or water and sweeteners are practical, as long as the products are correctly identified. Water, sugar, organic acids and low cost bulk juices are much cheaper than higher value fruit solids. Thus, unlabelled dilution and adulteration are common, unethical trade practices to be rigorously avoided and condemned.

Juice Safety

If there is any theme worthy of emphasis to the extreme in this text, it must be safety. Lapses in food safety do not only adversely impact the health of consumers. Flagrant safety errors have and will ruin the reputation and financial health of the offending food company. Food safety violations have caused the demise of companies and prison sentences for culpable executives. Worse yet, highly publicized safety lapses indict an entire industry, implicating the innocent majority together with the single guilty company.

The combination of fruit raw materials, food ingredients, juice preparation and processing methods, provide the technologist with an practically infinite range of possible juice products varying in quality, price and value. The liquid nature of juices and the availability of many natural and manufactured juice constituents offer a blending opportunity not available to formulators working with solid foods. The versatility of juices plus the many mixing options are also a temptation to cut corners by economic adulteration.

The ease with which juices can be altered with sugar, water or inferior juices has and continues to attract unethical suppliers. Fortunately, advancements in analytical chemistry and instrumentation together with fruit and juice composition data banks make adulteration easier to detect.

Detection methods have more than kept up with adulteration schemes. Detection of doctored juices is expensive but fairly straightforward. Thus large international movements of juice are reasonably well monitored, although local, marginal producers continue to slip though the net.

Such unethical practices can also be dangerous, as there have been many tragic incidences of poisonous ingredients being added to juices. Fortunately, most countries now have food regulations and enforcement mechanisms that prevent such deception. Despite these efforts, another serious safety problem associated with juice consumption has arisen. Contamination of juices with pathogenic microorganisms such as E-coli O157 H7 and salmonella have caused numerous illnesses and some fatalities. Although it appears that such incidences are increasing, in reality, surveillance and detection method are becoming much more sensitive. At the same time, outbreaks are becoming quite newsworthy. All reported cases of contamination by pathogenic microorganisms were due to fresh, unpasteurized juices. This is an increasingly popular segment of the juice industry, whose sales are now jeopardized by new regulations.

There is a curious irony in juice-induced food borne illnesses. Many common foods of plant and animal origin are much more hazardous and contain higher levels of more dangerous microbes than do fruits. However, if these foods are normally cooked before consumption, the inherently dangerous pathogens are eliminated. Fresh, unpasteurized juices receive no such treatment and, despite their health-promoting image, can harbour dangerous contaminants.

Acid fruit juices below pH 4.6 were once deemed a minor health threat due to the high acid. Furthermore, refrigeration temperatures (below 5°C) represented an additional hurdle to pathogen growth, until the discovery that Listeria monocytogenes can grow down to 2°C. When juice spoilage did occur, it was usually a reflection of the indigenous microflora, yeast, mould and lactic acid bacteria. Nonetheless, the emergence of hitherto unsuspected food pathogens with acid resistance combined with increase in susceptible individuals, immunocompromised, chronically ill, the very young and very elderly. More stringent regulatory safeguards are now called for at all levels, i.e. provincial, national and international. Thus for safety and economic reasons commercial fruit juices are subject to strict regulatory control. Safety must always take precedent and strict limits on production, harvest, transportation, storage, manufacture, processing, labelling and distribution now exist. These are incorporated into Good Agricultural Practices (GAPs) and Good Manufacturing Practices (GMPs) with Hazard Analysis and Critical Control Point (HACCP) procedures being applied throughout the food chain. These will be emphasized as appropriate.

QUALITY

After safety concerns comes economic and quality considerations. A product can be perfectly safe and completely fraudulent. As the sale of a product labelled fruit juices may in fact consist of no or little actual fruit components. Thus the development of standards for fruit juices is incorporated into most country's regulatory codes. These regulations include the processing employed, the amount of fruit content required for various juice designations, the soluble solids and acid levels, amount of added substances allowed such as sugar, acid, water, preservatives and reasonable sanitary standards.

In addition, grades may be assigned depending on the colour, flavour, consistency and absence of defects. Codex Alimentarius grades and standards are being promulgated and continue to be refined to provide a useful guide for standardizing national norms and facilitating intra and international trade. In the matter of a large volume juice trade, specific standards are negotiated between major suppliers and users. For example, the Florida Department of Citrus, the state's industry controlling agency, has standards that exceed other norms and prohibit citrus juice manufacturing practices common in other regions.

In addition to existing international, national and local fruit/juice standards, detailed, industry-specific written quality specifications should be developed. Along with the safety and regulatory aspects, clear raw material specifications facilitate efficient processing (no surprises on the production floor) and uniform, consistent quality to the wholesale and retail purchasers. Reliability and trust throughout the food chain is essential. These standards should consist of physical, compositional, organoleptic specifications, as well as limits for microbial and chemical contaminants.

Analyses can readily detect trace amounts of pesticides, industrial chemicals and vanishing low numbers of pathogens, some with "zero" tolerance levels. Thus there has been a substantial increase in import denials and product recalls based on levels of contaminants, microbial and chemical, that were undetectable and unimportant a decade ago. The economic loss, not to mention the liability and poor publicity, is adequate incentive for juice processors to consistently take regulatory matters very seriously.

Regulatory Concerns

In industrialized nations and international trade, every step of juice manufacture from raw material cultivation through marketing is subjected to some type of regulatory control. Although some regulations can be onerous, burdensome, even unnecessary, there is a definite need to control food commerce items. The history of food control and the situations existing prior to development and enforcement of food laws surely justifies governmental oversight. The two principal needs are:

- Insure the safety of all food materials, ingredients, processes and packaging.
- Safety is of paramount importance. In high throughput manufacture, if a safety hazard is not detected and corrected somewhere in the production/supply chain, many consumers can be adversely affected. We have seen how microbial or chemical contaminants (defects in a small amount of raw material) can be blended into a large batch to be rapidly and efficiently distributed and consumed over a wide region. Hence, mistakes have correspondingly greater consequences. Errors can affect large numbers or, in the case of undeclared allergens or low level hazards, only susceptible individuals, usually with more serious results. Neither is acceptable.
- Prevent economic fraud - inauthentic ingredients, mislabelling.
- Here human health is not compromised, but the consumer or parties in the supply chain are cheated. Furthermore, such cheaters have an economic advantage over honest merchants and, if not discovered and eliminated, thrive at the expense of others. In view of the complexity of juice and beverage labelling laws, there is a grey area between outright fraud and misstatements. A clever blend of high fructose corn, organic acids, flavours, colours, nutrients, and phytochemicals can possibly match any natural beverage in appeal and health benefit, but to claim it as such is clearly dishonest. Declaring the contents and accurate juice percentage (if any) on a label implying fruit juice content is legal.

Table 1. Some quality specifications for juice fruit.

Attribute	*Rationale*
Soluble solids (°Brix)	Defines juice strength
Titratable acidity and pH	Defines acid balance
Colour	Visual appeal
Freedom from defects–decay, insects/damage, mechanical injury, etc	Aesthetics, susceptibility to spoilage and contamination
Maturity	Optimum quality
Size, shape and uniformity	Ease of juicing
Flavour	Defines quality
Absence of pathogens, chemicals and extraneous matter	Defines safety
Low microbial load	Quality, shelf life

Raw Material for Juice

Initial considerations

As indicated, the sources of fruits that can be made into juices are nearly limitless in plants and the plant parts appropriate for juice. However, the fruit producer and juice processor must adhere to some very important guidelines in fruit selection and juice manufacture.

Fruit cultivar

Human selection of plants for their more desirable features has been practiced for thousands of years, initially by trial and error, more recently by scientific knowledge applied to plant genetics and molecular biology. Thus even within a given fruit there are numerous varieties available including some derived from the wild, others specifically developed by breeders and combinations of both. Recent advances in understanding the molecular basis of plant genetics and how the genes and growing environment influence traits, promises to revolutionize agriculture, once the politics and emotions are resolved.

Fruits originating in the wild have evolved to be reasonably robust and resistant to natural environmental stresses, diseases, insects and other predators. These traits, however, do not necessarily favour concentrated production systems or edibility. The exception is if consumption by wildlife promotes pollination or seed dispersal. Human population pressures further impact upon native plant species, therefore except for exotic plants remote from civilization, most fruits are substantially removed from their origin in time, genetics and location.

Special cultivars, developed with criteria such as yield, location adaptability, resistance to disease, insects, drought tolerance, harvest period, ease of cultivation and harvest, handling and storage durability, processing characteristics and sensory appeal (colour, taste, aroma, texture) are characteristics now determined by commercial breeding practices. Juice quality considerations are also important in many fruits such as citrus, grape, apple, pineapple and others. Specific cultivars dominate the juice trade, while others are preferred for the fresh market. Some cultivar end uses are dictated by fruit supply, condition and economics and varies by season, location and market depending upon fruit quality.

Nevertheless, all fruit species include cultivars that are highly inappropriate for juice (or consumption for that matter) such as bitter, intensely sour orange rootstock. Bitter watermelon contains cucurbitacum, a very bitter compound. Mango trees are usually grafted and chance seedlings range from those with deliciously sweet, smooth textured flesh to fibrous harsh terpeney fruit. Some fruit species are even toxic and avoided by experience. It is conceivable, even probable, that such inedible or toxic fruits may possess unrecognized nutraceutical value, but safety and palatability are the present criteria.

Cultivation practices

The condition of a fruit tree or plant throughout the season from flowering to fruit maturation can dramatically influence juice quality. Practically all cultivation parameters can influence quality and composition. Season, location, fertilizer application, form and timing, irrigation, amount of precipitation, temperature, sun exposure, plant spacing and pruning, disease, insect and predator stress influence fruit quality and yield. In the interest of quality, uniformity and yield, horticulturists constantly strive to overcome unfavourable conditions that may be a fresh, unexpected challenge each year. For the commercial grower there is never an "average" season or crop; quality, prices and supply vary accordingly.

Unforeseen, variations in weather can have a significant effect upon fruit quality, quantity and juice characteristics. Unseasonable freezes or hail can decimate a crop or send fruit destined for the fresh market immediately to the juice processor. Insect and disease damage also affects the fresh marketability of fruits and excess or inadequate rainfall influence juice solids, flavour and composition. In contrast to whole fruit quality, juice variability can be somewhat overcome by blending of juice stocks or amelioration, if permitted.

Cultivation practices also affect regulatory matters and labelling. For example, to label the product "organic" certain very specific criteria must be met, some based on perception more than scientific reality, but extremely important to proponents. As with juice purity regulations, organic standards must be rigorously adhered to as a matter of principle. Thus the use of fertilizer, pesticides, and all agricultural inputs, even timing and location relative to standard agricultural enterprises are carefully proscribed.

Cultivation also influences food safety, a topic for ongoing emphasis. In the case of fruit production, organic or otherwise, an additional concern is the use of manure as fertilizer. Food poisoning outbreaks involving juices have been traced back to the use of manure. Fruit that has fallen on the ground, becomes contaminated and is later collected and utilized. Even where manure is not used or is properly treated before application, the potential for fruit contamination by insects, wild life or field workers dictates cautious handling and sanitary practices.

Harvest season and maturity

The seasonal availability of fruit is the major determinate for juice manufacturing. Ideally, the maturation of desirable cultivars is either spread over a long period or different fruits can be harvested consecutively so that a juice operation can process for an extended season, if not most of the year. Although the northern and southern hemisphere seasons complement each other (and thus serve global fresh produce demands), the economics of producing juice from ripe fruit precludes such long distance hauling. The operations are best performed reasonably close to production areas. The globalization of the juice trade reflects regional availability.

There is an important distinction between plant maturation and fruit ripening. Maturity relates to the full natural development of the plant, whereas, ripeness is the desired stage for consumption or processing. Thus a fruit such as avocado or peach may be mature, fully developed and ready to be harvested, but far from ripe, requiring careful ripening conditions to achieve optimum quality.

Fruit are living structures and undergo complex changes during maturation, ripening and into senescence that influence quality. As fruits mature on the plant or after harvest the composition may change dramatically. Fruits are classified as (1) Climacteric, meaning they have a definite peak in respiratory rate for production of carbon dioxide and ethylene. (2) Nonclimacteric where there is a gradual decrease in respiration and CO_2 release.

Table 2. Factors influencing postharvest fruit ripening.

Factor	*Result*
Cultivar	Cultivar-specific response to holding conditions
High temperature	Accelerates ripening and microbial growth
Low temperature	Retards ripening, chill injury in some fruits
High oxygen (~21%)	Increased respiration and ripening rates
Low oxygen (<3%)	Promotes anaerobic respiration
High CO_2	Promotes anaerobic respiration
Ethylene	Accelerates ripening of climacteric fruit
Moisture (Rel. Humidity %)	>90%, microbial growth; < ~80%, wilting
Contamination	Losses during ripening
Physical damage	Speeds up ripening, decay

Whether these changes occur on the plant or post-harvest, it is important to understand and control them in order to optimize fruit and juice quality. For example, cold storage extends the shelf life of many fruits, such as apples and oranges, so the colder the better (avoiding freezing). In contrast, tomatoes and banana require storage temperatures above 7°C. Excessive holding at lower temperatures cause irreversible damage referred to as chill injury (higher temperatures are also detrimental). There are well-understood fruit and cultivar-specific holding and ripening regimens for all major fruit cultivars. With minor or exotic fruits, local knowledge and experience is the best guide; there is continual need for focused cultivar and location-specific ripening/holding research.

In general, fruit destined for juicing should be of good edible quality, full flavoured and substantially more mature than fresh market fruit. This favours softer tissues (more amenable to juice extraction and higher yield), higher sugar content, deeper colour and a lesser amount of acid. However, overripe fruit quality is inappropriate as the flavour and acidity may suffer.

Also, the fruit's defence mechanisms are diminishing, so incipient spoilage and microbial loads are apt to be higher. One of the major quality deficiencies in small juice operations is due to the use of under or over mature fruit, a propensity to use drops, culls and salvaged pieces.

Harvest and postharvest handling

At this stage the fruit have reached maturity and subject to a finite "harvest window" during which picking must take place. Cultivation and cultivar, as well as climate and weather affect this window. Heavy rains or extremes in temperature hasten or delay harvest and influence fruit quality. Harvest aids, machines and hand-pickers function better in moderate weather, but this isn't always possible. Delays and improper conditions extract a quality, yield and expense toll. Maturity standards as they influence colour, sugars, acid, fruit damage, etc., may be the basis of payment and dictate the end use, i.e. the fresh market, whole processing, juice, by-products, animal feed, or destruction (at supplier's expense).

Even given an acceptable, gentle, timely harvest, attention must be paid to field handling and transportation. Worker's judgement is critical in optimizing picked fruit quality as determined by size, shape, maturity and free from defects. Improper packing into field containers, rough terrain in transport and delays in transit can equally damage delicate or rugged fruit. A further consideration is the danger of contamination by field workers, containers and equipment. The inclusion of plant debris, mud, dust or soil contacting fruit can greatly complicate subsequent cleaning operations.

After cultivation there are many parameters impacting fruit and juice quality and safety. Some of these parameters are obvious, others are subtle while there can be complex interactions between several factors. Under most circumstances a fine balance must be achieved between these factors. It is worth emphasizing that extremes in any parameter are to be avoided. The proverbial "happy medium" is an attainable compromise, albeit species, cultivar, fruit condition and process specific.

General Juice Manufacture Principles

Unit Operations Applied to Juice

There are a number of unit operations involved in converting whole fruit to the desired juice, puree, or pulp product. Fruit handling depends upon the process design. If the raw material is destined for multiple uses such as fresh market, whole piece processing, additional products and juice, the flow scheme will differ greatly from one for a juice-only plant. In some circumstances, cleaning, sorting and inspection precede in-plant storage or the operations can be reversed or repeated immediately prior to juicing. As indicated, an ounce of prevention (preventing/removing contaminants from raw material) is worth a pound of cure (trying to clean up contaminated juice).

Table 3: Unit operation involved in juice manufacture.

Unit Operation	*Result*
Mass transfer	Fruit delivered, dry cleaned
Extraction	Washed
Separation	Sized, graded
Separation	Peeled, cored and deseed
Size reduction	Crushed, comminuted
Pressure application	Juice extracted
Separation	Solids screened
Deaeration	Oxygen removed
Centrifugation	Solids separated
Filtration	Clarification
Fluid flow	Juice transferred, pumped
Heat transfer	Enzymes inactivated, juice pasteurized and cooled
Concentration/evaporation	Volume reduction, stability
Mass transfer	Packaging, shipping

Normally, juice fruit does not receive the care in packing, transport or postharvest treatment reserved for fresh market and solid pack fruit. Operations such as cooling, washing, sorting and inspecting require attention to mass and heat transfer. Cooling depends upon heat transfer from fruit to air (possibly water). Cooling and cleaning can involve physical removal of surface debris by brushes or air jet separation prior to washing with water. These steps substantially decrease water use and speed up product flow. Efficient equipment minimizes cooling/heating energy and wash water use. When performed improperly the contamination level can actually be increased. Thus equipment and water sanitation is critical with chlorination and recycling usually necessary.

Pre-process storage

The logistics of production, harvest, transport and ripening dictate that many fruits must be held prior to juicing. The seasonal harvest window may be much shorter than the time required processing the entire annual crop and stabilizing the resulting juice and finished products. As with the aforementioned steps, careful holding is necessary to allow optimum ripening to occur or prevent spoilage and contamination. Producers and processors should have a good understanding of postharvest physiology, particularly as it relates to the species and cultivars in question.

With certain climacteric fruit, ripening and senescence can be controlled (delayed, maintained constant, accelerated, or promoted uniformly) by the judicious use of low temperature, moderate humidity and adjustment of oxygen, carbon dioxide and ethylene levels. These practices termed controlled or modified atmosphere storage can greatly extend the storage life of some fruits and vegetables.

With other fruit, holding is short term and serves only temporarily to prevent contamination and damage while accommodating processing flow. Some fruit can be frozen and stored for long periods. Freezing is costly, but promotes cellular breakdown and facilitates shipping and juicing. In hot climates where refrigeration or freezing facilities may be neither available nor practical for juice fruit, strategies such as just-in-time harvest, cooler night time harvest operations, rapid transportation and shady, well ventilated storage can help balance the processing load and stave off deterioration for a few critical hours. Many processors are disturbingly unaware of the importance of these procedures; but every little bit helps and the benefits add up.

Packing for transport should be gentle and sanitary. Simply throwing fruit into a bulk container and travel over rough terrain guarantees partial, unacceptable juicing during transit, prior to delivery. Moreover, if the delivery time is lengthy or if the lot must be held for more than a few hours at elevated temperature, incipient fermentation will follow. A major problem in handling and storage of fruit destined for juice is the relatively low value of the crop. Thus all post-harvest operations up to and including juicing often do not receive the attention they merit. Such produce failing to meet GAPs represents a safety as well as a quality hazard and is soon eliminated from trade in all but the poorest markets. Food safety regulations and quality standards continue to eliminate businesses employing such poor agricultural and manufacturing practices.

In some regions local processors stated this chauvinistic line, "We're doing fine, our products are inexpensive and are the consumer's only alternative." The result has been responsible for the dramatic increase in alternative beverages such as imported juices, carbonated soft drinks and other beverages. Most seriously, local fruit juices and all locally processed products have received an undeserved poor quality image.

Cleaning, sorting and inspection

If the fruit producer has followed GAPs and the harvest and handling have been accomplished effectively, fruit arriving at the processing plant should be reasonably sanitary and of optimum quality, thus simplifying succeeding operations. Nevertheless, quality cannot be taken for granted. In many cases the basis for payment is the condition of the fruit as received. Hence sampling and analysis for composition and quality are mandatory.

Table 4: Some quality criteria for juicing

Factor	*Criteria*	*Rationale*
Maturity	Ripeness	Optimum quality
Solids	Adequate level	Affects yield, flavour
Acidity	Appropriate pH level	Flavour, sugar/acid ratio
Colour	Fully developed	Juice appearance
Defects	Appropriate level	A few can be tolerated
Size/shape	Uniform	Ease of handling/juicing
Specific chemicals	Past analyses	Reflect handling/quality
Pesticide residues	Regulatory control	Legality of product
Foreign matter	Appropriate level	Reasonable limits
Microbial count	Low total, no or few pathogens	Safety/stability of juice
Aflatoxin level	Below proscribed limits	Juice safety

Government norms, an industry association, or agreement between producer and processor in advance of the crop purchase can dictate applicable quality standards. Some agreements are seasonal or even ready before planting or harvest. A representative sample of the shipment may be drawn according to statistical procedures, inspected for visible defects and foreign matter and then analyzed for microbial load, pathogens, pesticide residues, aflatoxin level, colour, sugar, acid, flavour, or other important safety and quality attributes.

Rejection at this stage is a serious matter, since considerable time, effort and expense has already been invested in the suspect lot. If the defect is correctable (cleanup, resorting and inspection) a price penalty may be imposed or alternate processing required. However, if the lot cannot be corrected, contamination (microbial or chemical) or far off-standard composition/quality disposal costs and reputation damages, result. In either circumstance process flow is disrupted and plant operations suffer.

Thus for reasons of quality and safety it is important that GAPs be in place before the harvest season, agreed to by all parties, followed strictly and documented throughout the production and delivery chain. At the juice processing facility, the GAPs interface with the processor's system (GMPs), which should be equally well designed and practiced. It is inefficient and illogical to take fruit that has been well handled up to this point and then subject it to inferior juice manufacture operations. The components of GAP and GMP systems are emphasized throughout this text.

Prior to juicing, the fruit can be washed, thoroughly inspected and sometimes sized (fruit-dependent). Inspection and removal of unsound fruit is very important, more so than in whole fruit processing. In solid packs one bad piece of fruit can cause a defect in one container, but after juicing that same piece of defective fruit can end up contaminating an entire lot of juice. In the same context, a few pieces containing microbial pathogens or toxic chemicals can and do raise havoc when juiced.

Dry pre-cleaning steps and water recycling systems may be required depending upon the availability and sanitary quality of water. GAPs should insure that dirty fruit is not delivered for processing. However, weather and delivery conditions may require the removal of dust, mud or transport-induced foreign matter. Water dumping/transport can serve to gently convey the fruit, but this is a dangerous practice, if dump water is not adequately chlorinated or otherwise treated. Maintenance of water quality is an expensive but essential feature of any quality juice processing facility. A screen shaker in conjunction with air jets and dry brushes can effectively reduce water usage, provided that the fruit can tolerate such mechanical handling.

Inspection can be manual, contingent upon workers observing and removing defects or automatic, effected by computer controlled sensors to detect off colour, shape or size. Sophisticated instrumentation operating at high speed is being increasingly employed in modern processing facilities, although the human eye, hand and mind still commonly make the final decision.

Crushing and juicing

There are many fruit-specific ways to extract juice. Some are well-established, large-scale procedures for commercial fruits and will be explained in detailed for those fruits. Operations range from kitchen to industrial scale depending upon volume, end use and raw material. The goal in juice manufacture is to remove as much of the desirable components from the fruit as possible without also extracting the undesirables. Thorough comminution maximizes the yield, but by so doing extracts substances from everything, i.e. seed, skin, core, etc.

Thus the compromise between juice yield and quality dictates the juicing and subsequent steps. Fruit with unpalatable skin and seeds must be treated more cautiously than one that can be completely pulverized. It is possible to minimize extraction of skin and seed components by a crushing regime that mashes or removes edible flesh, while sparing other portions. However, some fruit must be carefully peeled and deseeded or cored prior to juicing. Hand labour is the current alternative with many minor fruits, although there is an economic incentive to mechanize if possible.

Contour peelers, such as used with apples can be adapted to a range of fruits that are sufficiently firm and uniform to facilitate the rotary peeling action. A battery of analogous units was the precursor of the Ginaca, used successfully for pineapple. Peeling systems that are effective with some vegetables such as lye, abrasion, enzymatic, explosive, are less satisfactory with delicate fruit flesh. Nevertheless, cleverly designed machinery can greatly facilitate these labour-intensive operations. Under all circumstances, a final human inspection and piece selection/rejection step is mandatory.

Generally a whole fruit is more stable than the juice, unless rapidly preserved after extraction. So fruit should not be committed to juice until the material can rapidly be stabilized or the process go to completion. Attention to quality at the prejuicing step is extremely critical. Otherwise surface debris and portions of skin or seed can easily ruin the colour or flavour of an entire batch of juice. Of course, in a similar sense, microbial and chemical contaminants at juicing can and have caused disastrous public health consequences.

Juice extraction equipment ranges from hand operated crushers to tonnes/hour mechanical extractors. With soft or comminuted fruit a cone screw expresser or paddle pulper fitted with appropriate screens serves to separate the juice from particulate matter. Where skin or seed shattering is a problem, brush paddles can replace metal bars. Two pulpers in series with screens of ~1 to 0.2 mm can effectively clean up many juices and often yield a usable thick pulp by-product.

With thicker purees a pressing step may be needed and several options are available. For refractory material, pretreatment with a macerating enzyme with or without heating to ~60°C and holding up to ~40 minutes can greatly increase yield and subsequent pressing/clarification steps. Enzyme suppliers have selections of enzyme blends with hydrolytic activity against fibrous plant materials such as pectin, hemicellulose and cellulose to match the composition of most plant material.

Even without macerating enzyme addition, heating to ~70°C softens the fruit, inactivates native enzymes, reduces microbial load and affects an increase in juice yield. However, delicate flavours can be destroyed and unacceptable darkening due to enzymatic and non-enzymatic browning can occur. Rapid heating and cooling prior to juicing can overcome some of these quality problems. Traditionally mashed fruit and purees were batch heated to optimum macerating enzyme temperature (~55 to 60°C) in open steam-jacketed kettles with stirring. This step and subsequent cooling can be accomplished by swept surface heat exchangers or thermal screws (a hollow steam-heated auger in a steam-jacketed trough). Steam or cooling fluid flowing through both auger and trough effectively and continuously heats or cools the material. Pulpy, viscous puree cannot be passed through a plate or tube heat exchanger due to clogging problems, although, after solids reduction, these units are appropriate.

A thermal screw is a continuous system for both transporting and heating or cooling crushed fruit or whole small fruit. In a hot press regimen, the material passes directly from the crusher to the press holding tank via the screw, adjusted to deliver the crush at the proper temperature for enzyme activity, ~60°C. In continuous pressing, the screw can even be slowed down enough to accomplish the holding time and then feed directly to the press (or to another feed/holding screw). Only reasonably small particles can be so treated and care must be taken to insure even heating and avoid scorching of the crush.

Macerating enzymes also facilitate juice clarification steps. In some cases the naturally present enzymes (primarily pectinases) can be allowed to act at ambient temperature prior to pressing. However lengthy holding may favour spoilage organisms and the natural hydrolytic activity is apt to be slower than with added industrial enzymes. For fruits with delicate flavour or those where colour extraction from seed or skin is not desired, an immediate press at ambient temperature (referred to as a cold press) with or without enzymes is favoured. Juice yield will thereby be lowered; a compromise dictated to avoid undesirable heat-induced darkening.

A word of caution regarding commercial enzymes, enzyme preparations available in powder or liquid form, have a finite shelf life and should be refrigerated, since activity decreases rapidly at ambient temperature. These processing aids are fairly expensive and should be used sparingly. The solutions and powders are warranted only when the value of the additional juice yield and its quality exceeds enzyme cost. Also, they are mixtures containing individual enzymes, some of which may promote undesirable reactions, so consultation with suppliers and pilot testing is advised.

Another means of extracting water-soluble components from fruits and plant material is infusion extraction. The flesh is comminuted with added water to dissolve solids that are then separated from the pulp as described. Multiple extractions with temperature and pH adjustment and enzyme treatment can extract practically all-soluble solids. A counter-current flow where fresh solvent (usually water) contacts the last stage of spent pulp is quite efficient and used widely in the Citrus Industry. In this case, subsequent concentration of the extract is necessary to return to the initial fruit °Brix.

With soft, readily extractable fruits a steam extraction system has promise. An early example is a home extraction unit of Scandinavian origin. The fruit is in a strainer that drains concentrically away from an inner container with boiling water. Rising steam condenses in the fruit compartment or on the fruit, heating it and leaching out soluble that drain away from the pulp. Despite the use of high temperature, this system steam inactivates enzymes (blanches) and pasteurizes the fruit and excludes oxygen during extraction. Thus the juice has unusually good colour and flavour retention.

Independent of the system used, if the extract is then concentrated back to the original solids level, it is technically juice. However, labelling as juice depends on specific country regulations. The pulp or press residue of high value fruits can be extracted in this manner with water or other solvents to yield extracts containing pigments, nutrients, nutraceuticals, essences, or other useful by-products.

Pressing

The pressing operation can also range from manual to mechanical with complete automated systems common in the juice industry. Kitchen-scale juicers or food processors are effective for small quantities, but for larger multi-kilogram amounts, flow resistance and distance the expressed juice must travel to the press surface complicate pressing. The rack and cloth press increases the surface area to volume ratio and is quite effective, albeit labour intensive.

Other batch systems involve hydraulic basket presses where pistons move down into a press sack containing the crush. Better extraction is obtained by mixing the crush between multiple pressings to expose fresh surface near the cloth. This mixing step is accomplished in bladder presses by rotating a horizontal basket fitted with a chain or internal mechanism to redistribute the cake and reinflating the bladder to provide pressure on the crush. Vertical bladder presses inflated by air or water pressure are also in use. Periodic deflation of the bladder and manual cake redistribution is necessary to insure a fresh press surface.

At the other extreme are continuous screw presses capable of handling many tonnes/hour of crushed fruit. Care must be taken not to subject the press cake to excessive shear; or else seeds and other solids will contribute undesirable components to the juice. A more expensive continuous yet gentle system is a belt press where the pulp is pressed between porous belts by rollers. In most press configurations, adding several percent of a press aid to the crushed fruit can increase yield. Press aids consist of clean rice hulls or cellulose fibre that provide drain channels for the expressed juice. Multiple pressings or rotations (in bladder presses) further increase yield, but require more time and can extract undesirables if overdone. The press aid also serves to strain out particulates; juice expressed late in the cycle tends to be clearer (and darker, if browning has occurred). However, press aids can present a disposal problem, may be expensive and unless well refined, can contribute off flavours.

With some fruits allowing the crush to settle provides natural drainage. The pulpy fraction either floats or sinks for easy separation and greatly reduces the volume needing pressing. Fining with the use of bentonnesite or highly adsorbtive powder capable of flocculating colloidal material from the juice speeds up settling and is commonly used in wine clarification. Similarly, addition of protein or tannins as fining agents can not only remove suspended solid, but also form complexes with macromolecules that, if not removed, can cause haze formation in finished juice.

Juice clarification

For more fluid juices where cloud or turbidity is not acceptable primary extracted juice must be treated further. A settling step can help, if the juice can be held refrigerated for a few hours. At ambient tropical temperatures holding is not recommended. Rapid methods such as centrifugation and filtration can produce a clear juice. A continuous or a decanting centrifuge with automatic desludging to produce a clear or nearly clear juice is quite effective. The stream should be settled or coarse strained prior to centrifugation in order to reduce the sludge load in the feed going to the centrifuge. A fine mesh shaker screen can further remove particulates. A centrifuge is a very costly item; however, it greatly simplifies subsequent filtration steps and is an essential component in many juice processing operations.

Filtration

As with pre-centrifugation, the juice stream should be cleaned up as much as possible to reduce treated volume and increase throughput. There are many filtration systems well suited to various juices. These range from plate and frame filters, fitted with porous cellulose pads, to plastic, ceramic, or metal membranes. Diatomaceous earth mixed with the liquid serves to greatly increase the surface area and porosity of the filter bed and hence the particulate absorbing capacity of the filter.

In the extreme case a filter can have small enough pores to physically remove microorganisms from the juice (sterile filtration) or even remove macromolecules such as proteins and carbohydrate polymers (ultrafiltration).

These processing steps will be considered later. The production of a clear or brilliantly clear juice and the prevention of post filtration turbidity are the normal goals. Membrane clogging is a concern in any filter, so pretreatment to minimize particulates and flow patterns that self clean the membrane are essential.

The system can be made continuous by a rotary vacuum filter. A vacuum deposits fresh filter aid suspended in the juice on the drum surface. The juice passes into the rotating drum through the filter bed that is constantly renewed by scraping the spent material off the surface after a rotation. Various other filters use durable ceramic or metal porous membranes with backflush capabilities and flow patterns that minimize clogging. These systems operate in parallel to insure continuous operation and can reduce or eliminate the need for filter aid; a purchase and disposal expense.

Deaeration

In the operations described crushing, comminution, pressing, shaker separation, centrifugation and filtration, the fruit and juice are subjected to considerable aeration. The inclusion of oxygen can promote enzymatic browning, destroy nutrients, modify flavour and otherwise damage quality. Therefore, care should be taken to perform these steps rapidly, at low temperature and/or protect the material from oxygen, if possible. Sometimes preheating to inactivate natural and/or added enzymes is useful, provided rapid cooling follows.

Deaeration can be accomplished by either flashing the heated juice into a vacuum chamber or saturating the juice with an inert gas. Nitrogen or carbon dioxide is bubbled through the juice prior to storing under an inert atmosphere. Clearly, once air is removed or replaced by inert gas, the juice must be protected from the atmosphere in all subsequent processing steps. Deaeration, especially flashing off at high temperature, can also remove some desirable volatile aroma, another compromise facing the juice technologist.

A word of caution regarding processing systems

The excellent research reporting juice details - preparation, processing, composition, storage quality, etc. most likely came from academic or institutional research labs working with small, albeit representative samples of fruit. Handling is usually ideal with the major emphases on repeatability, speed and attention to detail. Rarely are the studies concerned with the logistics of handling large, i.e. commercial quantities, under industrial conditions.Thus scale-up and the attendant difficulties are not stressed. Just because a fruit can yield an outstanding juice in the lab doesn't mean it is commercially viable. Equipment from harvest to storage must be developed to efficiently perform the myriad of operations under varying and often vastly different and difficult operating conditions. Furthermore, these operations must match reasonably well the process flow scheme. Incompatibility in throughput, or highly sensitive procedures with excessive down time raise havoc with production flow and cannot be tolerated. Ingenuity and/or technology can eventually overcome these hurdles, but only with considerable patience and time.

Hand labour and care can represent a first step in scale-up and produce juice products for local niche markets. However, this is no substitute for well-designed mechanization, to replace or at least supplement human labour. Thousands of labourers cannot match several high-speed extractors for long. Some fruits mentioned here have superb eating quality, but will have scant juice (or solid pack) potential, until some degree of cost-reducing mechanization is introduced into the system.

Generally it is the inspection, cleaning, sorting peeling/coring/pulping operations that cause the greatest problem. This is a major challenge for small processors. If they don't continually strive to improve juice manufacture efficiency, somebody else will or the fruit will remain under exploited to the detriment of the growing region and consumers.

Curiously, the awesome ingenuity and craftsmanship demonstrated by third world mechanics in repairing and maintaining complicated machinery and equipment with little training or parts inventory has not been extended to food processing equipment design and development. Of course, all such equipment is kept running long after the anticipated design life, past zero depreciation. New and innovative mechanical solutions to tedious processing steps are either conspicuously absent or not shared within the local food industry. There is a critical need for such innovation, especially since existing (large scale/multinational) equipment manufacturers with the experience and talent view this as a niche market not worthy of the investment. Conversely, if appropriate small/medium scale equipment does exist, it is atrociously expensive and/or complicated to operate and maintain.

JUICE SPOILAGE

Freshly expressed juice, is highly susceptible to spoilage, in fact more so than whole fruit. Unprotected by skin or cell walls, fluid components are thoroughly mixed with air and microorganisms from the environment. Thus, unheated juice is subject to rapid microbial, enzymatic, chemical and physical deterioration. The goal of processing is to minimize these undesirable reactions while still maintaining and in some cases enhancing, the inherent quality of the starting fruit.

Table 5: Juice safety and deterioration hazards.

Hazard/Deterioration	*Result*
Microbial contamination	Survival/growth of pathogens
Microbial contamination	Rapid spoilage
Aflatoxins on fruit	Unsafe/illegal product
Pesticide residues	Unsafe/illegal product
Spurious dissolved matter	Unsafe, off-flavour
Spurious particulates	Unsafe, reduced quality
Enzymatic activity	Browning, consistency/flavour changes
Dissolved oxygen	Browning, nutrient and quality reduced
Metallic cations	Flavour/colour/nutrient losses, unsafe
Maillard reactants	Browning, quality loss
Extended holding	Quality deterioration
Colloidal instability	Sedimentation/precipitation/haze

The major concern must be microbial deterioration. The ubiquitous nature of microbes dictates that no matter how sanitary the raw material, juicing equipment and facilities are, the normal fruit microflora represents inevitable spoilage potential. The low pH of most juices favours yeast that rapidly initiates an incipient fermentation. Most civilizations have mastered this type of spoilage and directed it toward the production of alcoholic beverages. However, left uncontrolled it represents spoilage. Furthermore, the presence of aflatoxin producing mould contamination on the surface of fruits generally ends up in the juice.

Certain aflatoxins are potent liver carcinogens and commercial standards dictate parts per billion limits on aflatoxin levels, thus restricting the use of mouldy fruit. This is another valid reason for GAPs, sound harvest selection and inspection practices.

In fact, until recently the uncontrolled fermentation of improperly handled low pH juices was wasteful, but not deemed particularly dangerous. However, over the last decade fresh juice has increasingly been the source of serious food poisoning outbreaks and fatalities. Unpasteurized juice has been implicated in outbreaks of Salmonella and emerging pathogens such asE. coli O157 H7. These incidents have resulted in far stricter sanitary and labelling requirements for commercial fresh juice producers.

There are several reasons for increased fresh juice food poisoning incidents:

- greater use of manure as fertilizer for fruit crops, a particular problem for organically grown produce,
- increased demand for and consumption of fresh juice as compared to pasteurized and frozen concentrates. The "back to nature" movement promotes a strange philosophy regarding common food processing practices,
- a larger number of individuals are immuno-compromised and quite sensitive to low numbers of pathogens. The very young, very old, pregnant women, transplant patients and those with chronic disease conditions are extremely susceptible to food poisoning, compared to healthy individuals. The consequences of infection are correspondingly more serious,
- the pathogens are also becoming more robust and resistant to preservation techniques,
- efficient manufacturing and distribution systems insure widespread, prompt delivery and rapid consumption of juice products, both safe and tainted,
- detection techniques based on molecular biology can now rapidly identify very low numbers and "fingerprint" specific pathogens or their metabolic products,
- the news media is also more efficient in detecting and publicizing food sanitation failures.

Along with safety considerations, quality factors are also important. Sound fruit, reasonably free from microbial contaminants are subject to biochemical deterioration upon juicing. The mixing of fruit enzymes with substrate and air can rapidly initiate enzymatic browning. Plant phenols, polyphenol oxidase and oxygen react to darken many juices. There are many other enzymes active in juice capable of destroying ascorbic acid, modifying pectin and affecting colour, flavour and texture. So rapid processing and the use of heat or enzyme inhibitors are necessary with some juices.

Even barring microbial and enzymatic changes, other chemical reactions involving oxygen, metal cations and other juice constituents can occur to modify sensitive pigment, taste or aroma substances. There are literally hundreds of reactive compounds in the simplest of juices, so the reaction possibilities are enormous.

Another storage limiting factor affecting most juices is the common sugar-amine or Maillard reaction. Reducing sugars and amines, ubiquitous to plant cells and hence juice, slowly go through a series of steps to form brown pigments. In the baking of bread Maillard products are responsible for the desirable crust colour and flavour, but not in juices. Low temperature greatly retards Maillard browning, but does not stop completely these undesirable reactions.

Chemical contamination can also occur from the environment. The unauthorized or excessive use of pesticide chemicals is the most common and avoidable source, readily preventable by GAPs. Such contamination is particularly serious in organic juices where no pesticides are permitted. Even trace amounts of innocuous substances present in soil, water, or wind drift can cause rejection. Although the health hazard is trivial, the analytical sensitivity insures detection. Mistakes anywhere in the food chain are more serious. Unlabeled lethal white powders have been mistaken for food ingredients and added to juice, resulting in fatal poisonings.

The use of nonfood grade equipment in the processing line is a relatively minor safety concern that still impacts juice quality. Metals such as copper, bronze, aluminium, iron, galvanized, steel (except stainless) are easily attacked by fruit acids and, in turn contribute metal ions to the juice. These ions can have pro-oxidant properties and adversely affect flavour, colour, clarity and nutritive value. In addition, flexible connections and all food contacting surfaces should be stainless steel or food grade plastic, since off flavours and unauthorized plasticizers could leach into the juice. Certainly, contact with toxic metals such as lead, mercury, cadmium and zinc must be rigorously avoided. Food grade surfaces are significantly less available and more expensive than easily obtained "hardware store" alternatives, but much more durable, safe and ultimately cost-effective.

Physical changes can also take their toll on juice quality. Separation of dispersed particles can help or hinder appearance. Clear juices can turn cloudy or release an unsightly precipitate. Chemically or freeze-thaw induced colloidal reactions can affect juice viscosity to thicken or thin the consistency and influence the taste. In view of the numerous pathways to juice spoilage and delicate nature of juices, it is an impressive testimony to food science and technologies that juice stabilization techniques work as well as they do. Indeed, through most of history, the controlled spoilage (fermentation) to alcoholic beverages such as wine, was the tenuous but only preservation alternative.

JUICE STABILIZATION AND PRESERVATION

Preservation

Despite the many pathways to deterioration, there are a number of effective preservation methods that have evolved to combat spoilage. A principle tenant of food preservation is to maintain the quality and nutritional attributes while preventing spoilage. In general, the fresher the juice, the higher the quality, so the standard of excellence is often freshly prepared, unprocessed juice. As indicated, this is a very transitory product having a limited shelf life of hours or days even under the best of circumstances.

Refrigeration + sanitation

Two practical "processes" capable of extending storage are rigorous attention to good sanitation from production through juice preparation and low temperature holding. Even in the absence of pathogenic microbes, the natural microflora present will be active. The microbial load can be appreciably reduced by good handling and sanitary practices, but not eliminated. A holding temperature as close to the juice freezing point as possible (-1 to -3°C, depending on soluble solids), combined with efficient preparation and exceptional sanitation in preparation and packaging can extend storage life up to a month. Several accompanying difficulties are the possibility of other quality-reducing reactions comprising enzymatic, oxidative and the challenge of maintaining minimum refrigeration temperature throughout distribution, including in the consumer's refrigerator. Going from an exceptionally low refrigeration temperature to a merely good one (2°C to 5°C) can reduce shelf life from greater than a month to less than 3 weeks. One rule of thumb specifies each 10°C increase in temperature roughly doubles reaction rates.

Refrigeration combined with pasteurization and hermetic packaging can further increase storage life with minimum quality changes, but then the juice cannot be labelled "fresh". Thermal processing does away with the absolute need for refrigeration and inactivates enzymes. But heat-induced reactions change the character of the juice. With many juices heat is tolerable and commercially acceptable packs are available. However, there are some delicate-flavoured juices, most of tropical origin, that cannot tolerate even the most gentle pasteurization step. A major quality defect in pasteurized tropical juices and juice beverages is over heating or holding the product at too high a temperature before or after pasteurization. The resulting scorched flavour and colour does not do justice to the fruit so treated, easily reduces consumer appeal and destroys market potential.

Freezing

All preservation methods that allow juice storage above refrigeration temperatures, greater than about 5°C, are limited by Maillard browning and other slow but persistent reactions. Thus low temperature during manufacture and storage is a valuable hurdle to deterioration. Single strength juice freezing and storage in a low oxygen environment can maintain fresh character as well or better than any other process. However, it is a costly alternative and appropriate only where product value merits freezing and holding large volumes of single strength juice. Slow deterioration (over months) is exhibited by decline in nutrients, primarily loss of ascorbic acid, enzymatic activity as well as flavour, colour and viscosity changes. For frozen products, the closer to thawing temperature they become, the faster these reactions occur. Although the product cannot be labelled as such, subsequent thawing and proper packaging approximate fresh juice. "Fresh frozen" is the appropriate designation. The thawed juice then has a shelf life limited by sanitary conditions and pre-freezing treatments.

Combined methods

In circumstances where thermal processing is impractical, minimal processing employing hurdles is called for. The hurdle principle is based on the premise that while any single barrier to microbial growth may be inadequate for desired protection, a number of barriers together can enhance product stability.

Thus good sanitation is the first barrier to reduce microbial load; low storage temperature further retards growth. An acid environment of pH less than 4.5 (ideally less than 3.5) restricts the growth of many organisms. Fortunately most juices are acid, or amenable to acidification. The exclusion of oxygen is an additional barrier. (Although anaerobic conditions are dangerous with low acid foods and can even set the stage in high acid products for anaerobic spoilage organisms, i.e. yeast, lactic acid bacteria, etc.) Antimicrobial substances, either natural or chemical preservatives, also assist. Proper use of hurdles can appreciably lengthen shelf life of unpasteurized juices without unduly affecting quality.

Minimal processing is the strategy by which a number of hurdles are combined to effect near "fresh" juice characteristics. Although the sensory quality may approximate that of the unprocessed product, the term "fresh" cannot be used. The ubiquitous network of refrigerated and frozen production, storage and transportation facilities, including home units, encourages minimal processed juice products. Thus, even the slightly less fresh flavour exhibited by pasteurized juice is not acceptable to some. In developed countries fresh squeezed, refrigerated juices continue to capture sales from frozen concentrates despite a substantially higher price and shorter shelf life. Convenience and the closer-to-natural image apparently justify the 30 to 50 percent price premium and bulky storage requirements of these ~85 percent water drinks.

Competitive inhibition

A conceptually appealing process (actually a hurdle) is the inclusion of benign microorganisms into a food. These microbes, designed to grow well under conditions of storage abuse including temperature, oxygen level, pH, water activity, etc., will dominate spoilage and readily signal their presence thus harmlessly causing rejection of the spoiled item. In some cases competitive growth produces substances inhibitory to dangerous organisms, or affects a desirable chance in the food such as cheese ripening, alcohol production and lactic acid and flavour development.

Single strength beverages requiring refrigeration throughout the distribution chain. Indeed, the development of these fermented foods served to stabilize and improve the original materials. Taking the concept one step up the food chain suggests that competitive inhibition can be used to combat pathogens in food borne disease vectors i.e. insect droppings, animal manure/ intestine and even in the human body. Just as vaccination employing modified pathogens or their derivatives dramatically reduced infectious diseases, genetically modified organisms with tailored growth/inhibition characteristics will eventually protect minimally processed foods.

Thermal processing

The acid nature of most juices permits pasteurization, defined as the use of temperatures near 100ºC to effect destruction of spoilage organisms. Although spores conceivably can survive at a pH less than 4.6, outgrowth is unlikely. In contrast, at a pH greater than 4.6, spore heat resistance dictates a process temperature of greater than 115ºC for an extended time. Hence pH reduction by acid addition to turn low acid or marginal pH juices into high acid products is widely practiced.

A potentially effective process for fresh juice involves heating clean, well-sorted whole fruit for up to 1 minute at 80ºC. This greatly reduces surface contamination without influencing the underlying flesh that, if juiced in a sanitary manner, has the sensory quality of fresh juice. A great deal of research is underway combining surface heat with other decontamination practices. Whether these treatments will qualify as producing "fresh" juice for labelling purposes remains to be seen.

Canning

Standard canning procedures specify filling cans or jars with hot juice (~70 to 80ºC), sealing and processing at 100 to 105ºC for up to 10 minutes and cooling immediately. This is rarely done in a still (stationary or motionless) retort, since slow heating and cooling would ruin the quality. Instead, a continuous rotary retort provides rapid heating and cooling as a result of the juice being stirred inside the can by the headspace bubble movement during rotation. Another rapid system is the spin cooker/cooler where the spinning action provides good internal and external surface contact.

Hot fill

It is fairly easy to hot fill pack juices by rapidly heating the juice in a heat exchanger and filling containers with the hot juice measuring around 95ºC followed by sealing and inverting, thus pasteurizing the container. Reasonably rapid cooling is accomplished by rotary or spin action. This is known as flash pasteurization and can be achieved almost instantaneously. However, once in a container, cooling cannot be as rapid. The major quality problem is scorching, due to holding the juice hot, either before or after filling.

Many types of heat exchangers are used. The simplest is a coil submerged in boiling water with the juice flow adjusted to the desired pasteurization temperature. For pulp-containing juices or those likely to leave a film on the heating surface, a swept surface unit can continuously heat a juice stream. A plate heat exchanger with a regeneration section by which the cool entering product is preheated by the exiting hot product stream can be quite energy efficient, although the cooled stream must then be handled, filled and sealed in a sterile environment.

Hot fill has the additional advantages of driving air from the juice and ensuring a partial vacuum in the sealed container as vapour condenses upon cooling. Of course, deaeration and use of an inert gas during packaging are also useful. Sanitation is quite critical, since there are acid resistant moulds such as Bysochalmous fulva and Talarmyces flavus with unusually high temperature tolerance that require temperatures close to 100°C for up to 60 seconds for adequate pasteurization. Such a process could be excessive for some delicate-flavoured juices, thus requiring either gentler processes or exceptionally sanitary preparation. These moulds are not too common, but once established in a processing facility, they are difficult to eliminate.

Aseptic processing

A more elegant aseptic system rapidly heats, holds the hot juice to implement pasteurization and cools the juice before filling into sterile laminated paper/plastic/foil containers which are literally formed around a cylinder of flowing sterile juice. The rapid heating and cooling of the product guarantees microbial and enzyme destruction. A major advantage of aseptic packs is its adaptability to many size containers from single serving 200 ml portions to thousands of litre bulk packs. It is absolutely essential to pasteurize or in some cases sterilize the product, cool adequately and pack into sterile containers in a sterile environment, since the presence of a single viable organism can spoil an entire lot of juice. The evolution from simple canning to aseptic processing has greatly improved processed juice quality, but at a price.

Non Thermal and Other Methods

Sterile filtration

Another gentle and effective preservation method is sterile filtration. Clarified juice can be passed though a membrane filter with uniform pore size of less than 0.2 microns, thereby physically removing all microorganisms and virus. Of course, down stream sterility and aseptic packaging are essential and membrane leakage is disastrous. However, such juices have long storage potential provided endogenous enzymes are not active and reasonably low temperature storage is maintained. Juice where a cloud is normal or one that throws down a precipitate or forms a haze after filtration cannot be so treated. Experimental systems that separate the juice into a clear liquid, for sterile filtration and a pasteurized cloud portion then homogenizing and combining the sterile streams can be developed. Similarly, one strategy for juices containing considerable pulp is to separate pulp from fluid juice, process the juice rapidly in a heat exchanger and the pulp by slower means. The two streams are combined with homogenization to stabilize the blend and then packaged, all under aseptic conditions.

Chemical preservatives

Building on the hurdle principle, antimicrobials can effectively extend shelf life. Sulphur dioxide is quite effective in inhibiting both microbial growth and enzymatic and non-enzymatic browning and is standard practice in most wine making. A small number of individuals are quite sensitive to SO_2, so correct labelling and minimum levels are required. SO_2 is delivered as the gas or in the less noxious and more easily controlled form as potassium metabisulphite with about 60 percent by weight as SO_2. Dosage is reasonably self-limiting, since too little is ineffective and too much gives an unpleasant, pungent aroma. The preservative action is due to the undissociated H_2SO_3 molecule, so lower pH favours its use. Levels of 30 to 100 ppms, at a pH below 4.0 are effective. However, some asthmatics are so sensitive that simply being near SO_2 solutions is enough to trigger an adverse reaction, so caution is advised in use and labelling in foods if over 10 ppms is mandatory.

Table 6:Common juice preservatives and additives.

Ingredient	*Use*
Sulphur Dioxide	Retards microbial and enzymatic activity
Benzoates	Antimicrobial @ pH <4.5
Sorbates	Antimicrobial @ pH <6.5
Carbon Dioxide	pH reduction, anaerobic atmosphere
Ascorbic Acid	Retards enzymatic browning
Dimethylpyrocarbonate	Antimicrobial

Other preservatives such as benzoic acid, sorbic acid and to a lesser extent, CO_2 can be used individually or synergistically. Sodium benzoate and potassium sorbate are the preferred forms given greater solubility of the salts. Both function best below pH 4.0. U.S. Federal regulations limit benzoates to 0.1 percent and they are most effective against yeast and mould. Sorbic acid is effective over a broader range, up to pH 6.5. Benzoates and sorbates are often used together in combination with low temperatures to extend the shelf life of minimally processed juice drinks.

Although both sorbates and benzoates exist in nature, these so-called "chemical preservatives" are currently out of favour due to consumer misperceptions regarding their safety and the trend toward selecting "all natural" products. The term "no preservatives added" has the false connotation of improved quality. In reality, proper use of such preservatives can benefit safety and quality and prevent much waste, but is a hard sell in the marketplace.

There continues to be misuse of preservatives by some processors. Benzoates in particular are overused. Some pasteurized juices have a biting aftertaste due to excessive benzoate. The overuse is not harmful, but ironically completely unnecessary in adequately pasteurized juice. The philosophy "if a little is good, more is better" has no place in food processing. Under no circumstances should any preservative be a substitute for sanitary practices. Preservatives are only one of several hurdles that can be used to maintain quality and extend shelf life.

A preservation method that has been alluded to previously is alcoholic fermentation. This will be dealt with as a by-product, although in most cases it is the primary product. Winemaking is now far removed from simply "letting Nature take its course" by which naturally occurring wild yeast convert fruit sugars to alcohol. Low levels of ethanol represent another hurdle and levels of over 7 to 10 percent by volume partially inhibit other spoilage organisms. The addition of alcohol up to about 18 percent is a reasonably effective antimicrobial preservative, although the character of the juice is dramatically altered. Oxidation and other types of deterioration can still occur.

Weak fruit wine at about 6 percent alcohol is the starting point for the acetic acid fermentation and the manufacture of vinegar, which represents a value-added by-product of fruit juices. Under unsanitary conditions and in the presence of air, spoilt juice or wine can slowly turn to vinegar due to acetobacter contamination. And vinegar itself is a useful pH-reducing hurdle, although associated more with acidified vegetable products than fruits.

However, gross contamination of fruit juice is apt to result in a vile mixed fermentation with moulds growing on the surface, producing visible growth and a range of noxious organic chemicals, perhaps aflatoxins, with alcohol being produced at the bottom. How producers up until a century ago consistently got palatable wine or even vinegar from such natural fermentations is a tribute to their pre technology skills.

Newer methods

Unless fruit juices are handled with extreme care for fresh consumption or gently thermal processed and stored at moderate temperature, flavour suffers. There are a number of evolving preservation methods applicable to juice that do not require heat. For the most part these are experimental with processing details still under investigation. The cost of processing equipment and present engineering limits to continuous, high throughput systems are barriers not yet overcome. Nevertheless, the promise of shelf stable juices with freshly prepared characteristics is adequate incentive, so commercial systems will eventually evolve. Hyperbaric pressure

Subjecting juice to extremely high pressure in the order of hundreds of megapascals can destroy vegetative cells and some spores. Enzyme inactivation requires even higher pressures or longer exposure times, but relative stabilization can be achieved. Rapid pressurization cycles demand veryrugged, expensive equipment. Although continuous systems for liquids are under development, initial use will be applied to high value products.

High pressure using CO_2 provides the additional stress of reduced pH. In this case, lower pressures (~100 MPa) can accomplish greater microbial destruction and enzyme inactivation than higher pressures alone.

High intensity pulsed electric field

High voltage potential maintained in a flow chamber can pasteurize juice at only moderately elevated temperatures. However, enzyme activity is reduced to a much lesser extent than vegetative cells.

Oscillating magnetic fields

Magnetic fields are somewhat effective against microorganisms and easier to apply to fluids, but these are still in the preliminary experimental phase.

High intensity pulsed light radiation

A thin film of reasonably transparent juice can be subjected to intense light irradiation ranging from ultra-violet through near infrared as a pasteurization step. Such a process is practical only for water and clear juices. Shelf life can be extended provided that flow is uniform, no blocking particles interfere and the product is deaerated prior to treatment and refrigerated rapidly afterward. However, enzyme activity is not affected appreciably.

Irradiation

Ionizing irradiation has long been practical to extend the shelf life of fresh fruits. The process can be applied in juice manufacture by either irradiating the fruit prior to extraction or treating the juice. Although juice enzymes are quite resistant to irradiation, elimination of pathogens is accomplished by reasonably low doses in the order of 0.2 to 0.5 kGy. Interestingly, irradiation long scorned by technophobes with little understanding of radiation chemistry, is an effective way of safely maintaining fresh juice quality under refrigeration. Although the products must be identified as having been irradiated through the display of the international symbol the Radura, it should be popular as a safe alternative to thermal processing. Irradiation may have the more immediate use as a whole fruit surface pasteurization step prior to juicing. As an insect disinfestation process in whole fruit, it is unparalleled.

Concentration

There are a number of liquid concentration systems applicable to juice. Open atmosphere concentration of juices has long been practiced; simply boiling off water is at the expense of product quality due to heat damage. The process that has been most responsible for the wide availability, economy and popularity of fruit juices is vacuum concentration. Post war developments in the United States of America dramatically increased the demand for frozen concentrate, primarily citrus (orange and grapefruit), grape, apple and pineapple. Energy efficient multi-effect systems with essence recovery systems are widespread in major juice processing regions globally.

The application of vacuum concentration reduces the boiling point of juice and, when combined with short exposure to high evaporation temperatures, reduces heat damage. Process developments led to multiple effect low temperature evaporators operating at a maximum temperature of 50°C High thermal efficiency is achieved by using the vapour from the first effect as the heating media in the second effect and so on. However, enzyme activity required pasteurization at ~90°C. The thermally accelerated short time evaporation (TASTE) system that combined enzyme activation at around 100°C for several seconds improved concentrate quality at high throughput. Essence recovery from vapour from the first effect further advanced concentrate technology and is found in many industrial units.

Table 7: Juice concentration systems.

Method	*Advantages/Limitations*
Open pan boiling	Inexpensive, slow, low quality, energy intensive
Vacuum evaporation	Good quality, energy efficient, costly equipment
Freeze concentration	High quality, slow, costly equipment, limited solids attainable
Reverse osmosis	High quality, slow, costly equipment, limited solids, clear only
Electrodialysis	High quality, slow, costly equipment, limited solids, clear only

Since vacuum concentration strips the natural aroma from the juice, quality suffered. An initial solution with citrus was to over concentrate the juice to about 55°Brix and add back about 43 percent of fresh, high quality single strength juice to obtain 42°Brix and provide aroma. Essence recovery and add back was an improved solution. Although in the absence of recovered essence, juice added back is an effective solution.

Water removal as affected by concentration greatly simplifies juice handling, storage and shipping logistics. Some juices can be concentrated to over 70°Brix, thereby increasing the juice solid up to 10 fold. High Brix is limited by viscosity build up due to the presence of pectin substances and insoluble solids or pulp. In the concentrated and frozen or refrigerated form the juice can be held for extended periods, shipped, or stockpiled for future use. Indeed, the global trade in frozen concentrate has profoundly influenced juice and juice beverage developments, since long term stability and ease of transport make concentrates a readily available commodity.

Concentrates must be maintained cool, if not frozen to prevent quality loss primarily Maillard browning type reactions affecting colour and flavour. However, shelf stable concentrates of about 50°Brix are found in retail trade under conditions of good inventory control.

Jelly and Jam Manufacture

A considerable amount of fruit juice, puree, pulps and pieces are devoted to jellies and jams. Standard jelly or jam manufacture requires a minimum of 45 percent by weight fruit juice, pulp or pieces, sufficient pectin to set the gel and acid to reduce the pH to ~3.1.

Although traditional jelly making relied on the natural pectin present in fruits, commercially standardized pectin is invariably added in commercial operations. The process is related to concentration, as the juice is partially concentrated and sugar is added to increase the solids level to above 65 percent. Combined with pectin and acid and hot filled, jellies are reasonably shelf stable and, along with the pectin derived from certain fruits can be considered a co-product of juice manufacture. As with other concentration methods, the use of vacuum concentration during boil down and rapid filling and cooling, dramatically increases jelly quality. Quality and efficiency considerations now favour vacuum systems in large throughput operations, although open steam kettles suffice in small niche markets. The key to open kettle jelly manufacture is:

- Sanitary processing facility and equipment,
- Trained workers,
- High quality raw material such as fruit, sugar, pectin,
- Rapid, minimum heating to desired °Brix,
- Immediate hot filling into sanitized jars,
- Rapid cooling to less than 30°C,
- Low temperature storage, less than 30°C.

Although the combination of low pH, high solids and hot fill eliminates microbial spoilage, an opened container can be recontaminated by sugar-tolerant mould. Also, high temperatures during preparation or subsequent storage reduced quality. Ultimately Maillard darkening and off flavour development occurred. Very delicate fruits make acceptable jelly, but the flavour is far removed from the fresh or gently processed juice product.

Low calorie jelly is even more difficult to manufacture and store. Low methoxy pectin and calcium salts or other gel systems that are not dependent upon high sugar level provide adequate gel set. However, such systems are more susceptible to post process contamination and require refrigeration after opening.

Wine

Fruit juice, particularly grape, is closely associated with wine. Indeed, wine was originally spoiled fruit juice that over the millennia was appreciated and perfected long before juices were refined. Actually, the technical and lay literature devoted to wine far exceeds the total treatment of all juices and similar beverages combined. Wine is briefly mentioned here since it represents a high value added product from fruits and also a step in the manufacture of vinegar, a logical by-product of juice manufacture.

Crushed fruit or freshly expressed juice/puree is one step away from wine. The biquitous presence of all types of microbes, the sugar and nutrient content of fruits and generally low pH provides an ideal environment for "something" to grow. Airborne, that something will no doubt be mould, with an unsightly mass of mycelia and offensive odours. However, below the surface where oxygen is limited, yeast growth is likely, resulting in the production of ethyl alcohol and carbon dioxide gas. The classic Gay-Lussac equation is an over simplified but stociometrically reasonable depiction of the transformation of fermentable fruit sugars to ethanol and CO_2.

Wine manufacture is challenging, relatively straightforward and well treated in classic texts. Many of the steps inherent in juice production also apply to wine, including selection of fruit and attention to quality and sanitation.

Dehydration

The next step from concentrated juice is further removal of water to obtain a dry powder of less than 6 percent moisture. In the dry state ambient storage is possible, although low temperature still serves to extend storage life. Water is much easier to remove from fluids that can be pumped and manipulated through heat exchange and vapour removal systems. Mass and heat transfer becomes more of a challenge as the juice loses moisture and fluidity. There are a number of dehydration techniques applicable to juices. For economic reasons most are based on some degree of juice concentration above single strength, generally as high as practical, given the flow characteristic and mechanical properties of the concentrated juice and design features of the dryer.

Fruit juices are not as easy to dry as protein or starch based foods. The high sugar content, consisting primarily of the reducing sugar, glucose and fructose, presents a problem. At low moisture levels the products are quite hygroscopic, readily picking up moisture from the air to become sticky and difficulty to manipulate. Dried fruit products, especially juices must be protected from the atmosphere during and after drying. An ideal drying room is maintained around 10 percent relative humidity. The dry product is then rapidly packed into a hermetically sealed moisture barrier material, either a bulk container or final package.

Juice powder also presents a large surface area and is quite susceptible to oxidation without surface bound water to protect reactive sites. Rapid destruction of ascorbic acids and carotenoids occurs, unless oxygen is excluded during packaging. A nitrogen flush may help retain low oxygen conditions in the package, sometimes assisted by an internal oxygen-absorbing sachet or material built into the package inner surface.

Another difficulty is the sticky point of the dried juices. Even with little remaining moisture at high temperatures (greater than ~ 70°C), the product will be in a glassy syrupy state and stick to machinery, vastly complicating materials handling. Malto-dextrans or other low dextrose equivalent (DE) corn syrups can reduce the sticky point, but then the juice powder is not pure. Means of overcoming these limitations are described for distinct types of drying. Nevertheless, equipment modifications and more demanding product handling procedures are required to overcome stickiness and the hygroscopic nature of dehydrated juices.

The simplest dry product is fruit leather or fruit rolls, obtained by air drying a thin layer of juice, pulp, or concentrate on a flat surface. Gentle low temperature drying in a pan yields a plastic, sticky sheet about 1 to 3 mm thick of the original product containing less than 15 percent moisture. Depending upon the sugar profile, temperature and ambient humidity, the leather will range from slightly to moderately sticky to the touch. Sucrose, low dextrose equivalent (DE) corn syrup or other humectants (glycerol and other sugar alcohols) can be added to reduce the hygroscopic nature. Increasing pulp content provides chewiness to the roll. Ideal moisture content is around 12 percent; over 18 percent the sheet can support mould growth and below about 10 percent the sheet is too brittle.

Vacuum Drying

As in evaporation processes, the juices can be dehydrated under vacuum to reduce drying temperature and enhance quality. Continuous and batch systems are in operation. For economy and quality reasons, the juice is usually concentrated prior to vacuum drying. If the consistency is thick enough the concentrate can be puffed by applying rapid vacuum and maintained in this state to produce a puff-dried solid. The puffing produces an expanded surface area that facilitates moisture diffusion.

Foam mat drying accomplishes the same effect without vacuum. A foaming agent is added to the concentrate with whipping to produce heat stable foam. Gentle air-drying results in dry foam. A continuous system applying a foam mat and blowing holes in the mat prior to drying has been developed. The puff or foam mat dried material is then ground to a readily reconstitutable powder. Foam drying has the advantage of using simple air drying, but the disadvantage of producing a large volume of foam to obtain a relatively small amount of dried juice powder, requiring protection from ambient humidity.

The more efficient high throughput air dehydration techniques such as spray drying and drum drying are marginally applicable to juices. Aside from adding malto-dextrans, a spray-drying pattern must be designed so that the juice spray dries and is cooled below the sticky point before the particle hits the dryer wall and enter the collection system. A circuitous flow pattern and the introduction of a cooling zone accomplish this. The cool, dry particles do not stick but flow and are packaged as a powder.

Another system employs a high tower in which spray dried material at the top has sufficient time to drift downward and cool before hitting a surface. Drum drying employing a hot roll rotating in the fluid or fed from between rollers. The film adhering to the roller surface is rapidly dried as it rotates out of the feed tank. Although the dried juice is in the glassy, sticky state, cooling and grinding under low humidity produces a juice powder. Vacuum drum drying in which the dryer in enclosed in a vacuum chamber can reduce heat exposure and affect a puffing of the juice on the drum.

The gentlest and most expensive dehydration method is freeze-drying. The juice or concentrate (at about 50 °Brix, the highest practical to maintain the frozen state) is applied to the belt, the vacuum puffs the juice slightly and it is then frozen and subjected to a vacuum well below the triple point of water, 4.6mm. Heat, applied by conduction from the shelf, infra-red radiation, or even microwave results in sublimation of the water in the frozen juice to vapour which refreezes on a condenser coil in the dryer. Since the juice never thaws, it dries as a porous sheet with minimal exposure to high temperature. If the vacuum is replaced by nitrogen gas and the dried juice layer is ground, transferred and packed hermetically under low humidity and oxygen pressure, optimum quality results.

Foods for astronauts (space food rations, including juices) are often freeze dried to reduce precious weight and volume. The freeze-dried juice is compressed to reduce volume and packaged under inert atmosphere in a pouch that is oxygen and moisture impermeable and designed to facilitate rehydration. The juice is rehydrated by injecting into the pouch water generated from the fuel cell reaction of hydrogen and oxygen, so the water "paid its way" into space. Freeze-dried juices are extremely expensive, since water removal costs exceeds US$2/Kg of water removed. The process is quite functional for military, space and backpacking applications where saving weight and space is so critical. With adequate protection from high temperatures, moisture and oxygen reconstituted juice quality is high and a shelf life of several years or more can be accomplished.

Juice Packaging Systems

Ever since early humans gathered semi-solid or liquid provisions, the problem of transporting and storing fluids has needed to be addressed. In clever adaptations to environmentally available materials the problem has been solved, first with animal skins/entrails and tightly woven or shaped/carved plant matter, then by industrial manufacture, i.e. pottery, glass, metal..

Parallel with the development of juice processing systems has been accompanying advances in packaging technology. The 3 piece can and glass-canning jar can be replaced in some applications with 2 piece aluminium cans, light weigh shock resistant glass, plastic and laminate materials. All facilitate greater efficiencies such as rapid throughput, product protection, lower cost and consumer convenience.

The impermeable and inert nature of glass, combined with transparency makes it the package of choice for sales appeal and upper price beverage packaging. Although brittle and relatively heavy, glass containers have visual appeal and reuse potential. Technical innovation has greatly reduced wall thickness, weight and thermal shock susceptibility. Nevertheless, hot filled or processed glass must be carefully cooled to prevent shattering. Brown and green tints limit light sensitive reactions in the product. However, the visual quality of some juices dictates a clear wall, at the expense of light-induced deterioration for shelf stable, pasteurized products.

These same pasteurized juices are found in metal cans. Initially tin plated steel with or without enamel linings was used. In some cases the tin contributed an expected flavour, otherwise, the enamel protected the tin from contacting the juice. Enamels consisting of vinyls and epoxies protected the tin or tin free steel inner surface. Enamel-lined aluminium cans, due to lightweight and ease of manufacture, are replacing tin coatings. Despite rugged enamels, some juices blends and products containing nitrates or SO2 are highly corrosive, so care must be taken to establish liner/product compatibility under realistic storage conditions. Three-piece tin plate cans from 200 to 1 300 ml are still in use for juices, but increasingly, PET (polyethylene terephalate polyester) bottles predominate in retail markets. There is currently rough price parity between glass and PET container. Energy costs are higher for glass and raw materials costs more for plastic, although PET has an edge for larger containers. Moreover, recent technological innovations, such as the incorporation of acetylene gas into PET and polyethylene naphthalate (PEN) or PEN-PET blend and copolymers promise enhanced barrier properties, light protection and printability.

For processes where the container is not subjected to pasteurizing temperatures, 80 to 95°C, light weight, durable plastic/paperboard/metal foil laminates are common. Aseptic processing lines literally surround a pasteurized and cooled juice stream with a sterile

package and seal the juice in segments of tetrahedral or rectangular shape. Sterile filtered juice can be handled similarly. Aseptic cartons consist of about 70 percent paper, providing strength and form; 25 percent low density polyethylene, for sealing and moisture barrier; and 5 percent aluminium foil, as the light and oxygen barrier. Container roll stock is sterilized during the forming step by hot hydrogen peroxide. Retail packs of 250 ml include a plastic straw and orifice space for puncturing by the straw's point.

Since there are plastic containers capable of withstanding pasteurization temperatures available, hot fill or post fill pasteurization is now common. However, in regions where such technology is impractical, juices are filled into all types of inexpensive plastics, including unpasteurized fresh juice, juice beverages and imitations in polyethylene bags for immediate consumption. When dosed with high levels of benzoates shelf life is extended at the expense of content quality.

- *Ease of recycling* - Glass, metal and some plastics have this feature. Indeed, some governments impose a recycling tax on packaging, reflecting the disposal cost of the residual material. Thus some laminates and composites are placed at a disadvantage; durability in service extends to the landfill. Of course, in less industrialized regions used, durable bulk and retail juice containers are held in high regard and find many domestic and industrial uses,
- *Uniformity of bulk packaging* - In the handling of pallets from shipping to retail the size and shape of over wraps is critical. Some shippers and retailers now demand standardization. This facilitates mixed pallet loads, loading and unloading of cartons and display case set up.

Storage requirements

In juices treated by any of the fore mentioned processes that destroy, remove, or inhibit microbes, there is still the potential for deterioration. This can take the form of intrinsic spoilage due to juice constituents or extrinsic induced by the environment. If residual enzyme activity remains, colour, flavour and consistency changes can occur.

Extrinsic changes can be minimized, if not prevented by proper packaging. Juice protected from oxygen, heavy metals and light is quite stable, except from the most unavoidable deterioration, a combination of the intrinsic such as juice chemistry, the Maillard reaction and the extrinsic such as elevated storage temperature. Sugar-amine browning accompanied by other degradations occurs slowly at ambient temperatures (20 to 25°C) but rapidly above ~30°C. In warm climates or under exposure to improper storage conditions, juices deteriorate within a few days. A closed truck sitting in the sun easily reaches 50°C, to the detriment of any food material inside.

Active packaging

The incorporation of oxygen absorbing, metal chelating, or antimicrobial agents as an insert or built into the packaging material is an effective strategy. Although of greatest value in respiring fresh produce, employing "active packaging" can protect food and minimize or retard inevitable deteriorations; another example of the hurdle principle in action.

Product identification

One important role of packaging is to provide information. Traditionally a label was required for the consumer, but the manufacturer and all others in the distribution chain, including retail sales need now identifying data. The introduction of the Universal Product Code (UPC), those small bar code strips affixed to most retail items from food to machine screws and laser scanned at checkout counters, has revolutionized marketing.

Along with price identification, UPC provided almost instantaneous information on sales location and volume, inventory and details relating to promotion, pricing and competitive products. The next generation promises even more information-rich devices and ease of use. Smart Packaging (e-PC) employs a packaging tag either emitting a radio frequency or having an electromagnetic charge for remote sensing, with orders of magnitude more information storage capacity of value to all in the juice distribution chain from fruit grower to juice consumer and even beyond to the package recycler.

Bulk storage

Juice manufacture and utilization logistics dictate that vast quantities of juices must often be seasonally held and/or transported long distances. The product must be completely stabilized for such handling. In the case of single strength or low Brix concentrates capable of tolerating thermal processing, aseptic processing is one alternative. Under aseptic high temperature-short time (HTST) processing conditions heat damage is minimum. Provided that the system down stream from the pasteurizer is sterile, including cooler, filler and container, many hundreds to thousands of litres can be processed, transported and held effectively.

Under reasonable sanitary conditions, for instance low microbial load, juices can be maintained at low temperature in a carbon dioxide atmosphere. High CO_2 pressure is more effective, but required reinforced storage tanks.

For more delicate products, single strength juice can be frozen practically unchanged, except for the expense and efforts required to freeze, store and eventually thaw large volumes of juice. Deaeration and rapid handling minimize enzymatic deterioration.Frozen concentrate at 68 to 72 °Brix held at less than -20°C is widely practiced. The product can be warmed slightly for pumping and has a lengthy storage life. Freezing and storage requirements are proportionately less than for single strength or low Brix concentrates. Since bulk storage at cool temperature is costly, inventory control and utilization scheduling are important to minimize holding time. However, bulk storage can be a useful hedge against crop shortages and/or fruit price fluctuations.

JUICE AND BEVERAGE BLENDS

Despite the strong appeal and tradition that many pure fruit juices have, there are logical reason for producing single fruit and mixed pure juice blends and juice products containing less than 100 percent juice. These are:

- Maintaining a single juice's uniformity within and between seasons by blending multiple cultivars to insure a consistent product,
- Blending of different juices to overcome the high cost of some juices,
- Overcoming scarcity and/or seasonal availability of certain juice components,
- Balancing out excessively strong flavours, primarily high acidity, astringency, or bitterness,
- Correcting low soluble solids level,
- Balancing juices with weak or bland flavour, but possessing other positive
- attributes,
- Improving poor colour or colour stability, of otherwise desirable juices attributes,
- Balancing extremely good, stable colour, with other positive attributes,
- Emphasizing unique nutritional or phytochemical properties,
- Overcoming undesirable single strength juice consistency.

Blending offers the opportunity to adjust sugar/acid ratios and compensate for other imbalances in juice from a single harvest or cultivar, since many factors influence the composition and quality of juice. These differences can be dramatic and complicate the task of a manufacturer committed to a uniform product with established standards. Fortunately, by blending several batches of juice with complementary compositions a uniform, standard juice is practical. Adjusting 100 percent juices is much more of a challenge than manipulating acid and sugar in juice beverage blends.

In a similar sense defects in many juice quality or nutritional attributes can be overcome by proper combination of juices. Further adjustments call for additional ingredients. Extremely acidic and/or strong flavoured juices completely mask more subtle juices. In which case, non-juice sweeteners can greatly extend the juice, although labelling regulations must be precisely followed.

Blending Strategies

One of the primary considerations in choosing individual components and preparing juice blends is °Brix/Acid ratio (°B/A). Depending upon the juices involved this ratio determines the sugar acid balance and influences the perception of each. For example, a 10 percent solution of sucrose is moderately sweet. Adding 1 percent citric acid for a °Brix/A of 10 produces an intensely sour sensation. This solution requires a few percent more sugar before the sensation of sweetness is once again dominant. Certainly, juices are much more complex but °Brix/A remains an important factor.

Even at optimum maturity, °Brix and acid can vary appreciably. Recent cultivar selection and breeding efforts have decreased acid levels and increased solids (sugars) in some fruits to facilitate fresh eating and juicing. In addition, there are consumer preferences to consider. Orange juice at a °Brix/A of 12 is standard in North America, while in the tropics 15 or more is deemed acceptable. Although a given °Brix/A can be obtained by modifying either Brix or acid (or both) juice blends or beverages with less than about 7°Brix are deemed weak and watery. Those containing more than ~1.2 percent acid are sour, independent of °Brix/A.

Added sweetener may include cane or beet sugar (sucrose), high dextrose equivalent corn syrup, or high fructose corn syrup. The syrups have a cost advantage over sugar in the United States of America. Non or low caloric sweeteners also have a place in juice blending. In this case the °Brix/A is inappropriate and the perceived sweetness must be balanced by trial and error, since the artificial sweeteners contribute little soluble solids and have varying characteristics depending upon the juice mix.

An additional concern is the organic acid used to adjust the °Brix/A. Normally the predominant acid naturally present is the best choice, i.e. citric for citrus, malic for apples, tartaric and/or malic for grape. Acetic and lactic which are characteristically associated with fermented vegetable products are rarely used exclusively, although the former along with minor acids such as adipic, succinic, shikimik may contribute to an acid blend.

At one time inexpensive bulk juices were deacidified by ion exchange with the colour and flavour removed by activated charcoal. The treated juices were concentrated (removing any surviving aroma) and added simply as sweetener, proclaiming the product as pure 100 percent juice. However, since the treatment essentially removed all original juice character, such material now cannot be labelled as juice, only as natural fruit sugars. These are suitable in "all natural" blends. An ion exchanged "reduced acid" orange juice has recently been introduced to the US market and bears the label '100 percent juice'.

Aside from adjusting flavour, colour, viscosity defects and correcting juice composition, blending can employ juice as the vehicle for delivery of important nutrients and phytochemicals. Pure juices or blends are routinely fortified with vitamin C and more recently calcium. In view of nutrition surveys indicating low calcium levels in the US diet, there has been a proliferation of calcium addition to foods and beverages.

In Japan foods for special health use (FOSHU) are an established category and the 1994 United States Dietary Supplement Health and Education Act (DSHEA) opened the door to dietary supplements. Juices and juice beverages are ideal and popular carriers for both recognized and fringe supplements. In fact, many natural components of juice are now recognized as having phytochemical value and are aggressively promoted in their own right.

Blending procedures require careful economic considerations. Depending upon year, location, weather and other factors, juice prices (usually valued as concentrate) can vary by a factor of 2 to 3 over a season. The more desirable and relatively scarce juices such as cranberry, raspberry, blueberry and lesser-known tropicals always have a high price, sometimes up to 100 times more than common juices. Bulk concentrate supplies of generic apple, pear, grape, orange and pineapples are usually readily available in the global market and relatively inexpensive (although premium cultivars demand higher prices). Such juices can serve as a logical blending base for other juices and added ingredients, for example sweeteners, acids, nutrients, preservatives, flavours, colours, etc. It is then up to the juice technologist to creatively blend in more costly juices to achieve the required quality and image. The regulatory aspects are quite important, since all ingredients must be properly identified on the label, along with the total juice percent quantified on a single strength basis. Additionally, there is legislative pressure in some industrialized countries with strong domestic agricultural interests to demand "country of origin" labelling of fresh produce and ingredients, including juice.

The economic incentives for blending are persuasive and require careful label reading as well as understanding by the consumer. This is not easy, since juice percentage declarations are inconspicuously printed, compared to promotional material. There is no question when a beverage is 100 percent juice as that fact is conspicuously displayed on the label. Also, a beverage can have juice as the first ingredient (present in greatest amount) on a single strength basis and still not be the major ingredient on a solids basis. High fructose corn syrup or other added sugars are reported on a concentrated syrup or dry weight basis, thus representing considerably more solids than the juice. Even orange juice with 5 percent sugar added (a common practice in some countries) is actually diluted ~30 percent on a solids basis.

An average supermarket carries several dozen juices and juice beverages in many sizes, shapes and forms including frozen, refrigerated, shelf stable single strength, concentrate and dried powder. Items range from 100 percent juice to completely artificial, perhaps with some added natural flavourings and colours. The price, single strength equivalent, can vary over 5 fold. Not surprisingly, "100 percent Natural" or "100 percent+" (of specified nutrients) on the label can contain little, if any juice.

Nevertheless, there is a certain health and image aura about 100 percent juice products that can be maintained and encouraged by judicious blending of complementary juices and accompanying marketing skills. "Natural" is an effective sales tool, even when the juice technologist can and does improve upon Nature.

Complementary Products and By-Products

A commendable trend in fruit processing has been the increased attention to waste reduction, resource conservation and by-product utilization. Driven by environmental regulations, economic incentives and the cost of energy and waste disposal, all food producers must now pay more attention to by-products. Moreover, many juicing operations evolved from the need to dispose of materials generated during fresh market and solid pack fruit processing accompanied by the realization that fruit residuals had juice potential.

In addition, the current emphasis on "natural" products is having a profound effect on the demand for fruit ingredients. Although often more costly, difficult to isolate and less functional (in stability, potency and activity), natural ingredients have a much greater market appeal than commonly used (functional and safe) synthetic analogs that are now considered "artificial" and "unnatural".

Pectin

Citrus, apple and a number of other fruits have sufficient pectin to merit recovery from the by-products of juicing operations.

Essential Oils and Essences

The taste and aroma appeal of most fruits is readily obtained in concentrated form by various extraction and/or distillation techniques from peel, flesh, juice, concentrate or combinations. The classic example is citrus.

Pigments

The natural pigments in fruits are also in demand and highly pigmented fruit cultivars are the focus of breeding efforts. Natural pigments are far less stable than synthetic colourants. Anthocyanin colour is highly pH-dependent as some range from bright red, pH less than 3.0 to green at pH greater than 5.0. Carotenoids are susceptible to oxidation, particularly in dry juice powders.

Phytochemicals

The last decade has seen an amazing paradigm shift during which the potential health value of food components, beyond common nutrients is emphasized. These substances, termed nutraceuticals in plant foods and phytochemicals, are the focus of a whole new industry. Fortuitously, fruits and fruit juices play a major role both as dietary sources and carriers of nutraceuticals. Actually, the field is in its infancy. Probably a number of hitherto unrecognized valuable substances could very well be discovered in juices and fruit extracts.

Without a doubt, the reverse applies for potentially dangerous substances, excess amounts of beneficial components or hazardous interactions are likewise possible, even probable. Nevertheless, increased research on the health value of juices promises to enhance the appeal and value of traditional and exotic juice products plus by-products.

Fruit by-products have long had economic value. Improved recovery technology, recognition of valuable components combined with stringent disposal regulations (and cost) provides the economic incentive to make a valuable resource from waste. In this regard, comparatively unknown tropical plants have an intriguing future, hopefully exploited by and in the countries of origin.

References

Ashurst, P.R. (Editor). 1995. *Production and Packaging of Non-Carbonated Fruit Juices and Fruit Beverages*. Blackie Academic and Professional.

Chen, C.S., Shaw, P.E. and Parish, M.E. 1993. Chapter 5 *In* Nagy, S., C.S. Chen, & P.E. Shaw, eds. *Fruit Juice Processing Technology*. AgScience, Inc., Auburndale, Florida.

Downing, D.L. 1972. *Homemade Fruit Juice Press*. Special Report No. 8. N.Y. State Agric. Expt. Sta., Geneva, NY.

Nagy, S., Chen, C.S. and P.E. Shaw. 1993. *Fruit Juice Processing Technology*. AgScience, Inc., Auburndale, Florida.

Tressler, D.K. and Joslyn, M.A. (Editors). 1961. *Fruit and Vegetable Juice Processing Technology*. 1st edition. AVI Publishing Co. Westport, CT.

6

Manufacturing Jams and Jellies

Fruits have been part of man's food since times immemorial. Among the preserved fruits, jams and jel-liesconstituteimportant productsand afford means of utilizing alarge amount ofsound cull fruits unsuitable for other purposes. It is an older method than canning and freezing. Historically, jams and jellies may have originated as an early effort to preserve fruits for consumption in the off-season. As sugar manufacture became more affordable, the popularity and availability of these fruit products increased. Jams in their various forms are probably the easiest by-products made of citrus fruits. The earliest published record of jelly making appeared in the later part of the 18th century.

The preparation of jams and jellies was developed as an art by the housewife, and served as a means of preserving fruit, corresponding to the time, when the fruits were being harvested. Science as applied to jam manufacture seems often to be regarded by the general public as synonymous with sophistication. Factory-made jam must conform to certain specifications and standards, not essential for a home-made jam, for instance; it must be of a consistency firm enough to meet the demands of confectioneries and to withstand handling during transport.

The consistency depends upon the presence of pectin. Hence, scientific jam manufacture is largely based on the correct application of knowledge about pectin and the laws governing the formation of the pectin–sugar–acid gel. Jam and jelly products were prepared with a high concentration of dissolved solids so that fermentation could not occur. However, only pectin and sugar are not sufficient for the formation of the products. Equally important is the acidity of the fruit, resulting in a definite equilibrium in the "pectin–acid–sugar" system.

Pectin is the most essential thing in the formation of jam and jelly. Lack of knowledge about the requirements necessary for the pectin gel formation, frequently contributesto products of undesirable consistency. The preservation industry awaited the accumulation of sufficient information on the chemistry of pectin and its gels to control the consistency under commercial conditions. As the knowledge of pectin chemistry increased, production grew and to some extent replaced home-prepared products. Consequently, the jam and jelly industry assumed considerable magnitude.

Ingredients for Jams and Jellies

Definition and standards of identity for various fruit preserves and jellies have been issued by the Food and Drug Administration under the Food, Drug and Cosmetic Act under which ingredients for jams and jellies have been discussed. A brief outline of these standards has been presented here. The U.S. standards for Grades were established by the production and marketing administration.

Jellies are viscous or semi-solid foods made from a mixture of not less than 45 parts by weight of saccharine ingredient. The mixture is concentrated by heat to such a point that the soluble solids content of the finished jelly is not less than 65%. Spices, sodium citrate, sodium potassium tartrate, sodium benzoate, benzoic acid, mint flavor, and harmless artificial green coloring may be optional ingredients. Optimal saccharine ingredients are corn sugar, invert sugar syrup, sucrose, honey, or combinations of these. Pectin and designated organic acids may be added to compensate for deficiencies of these substances in fruit juice. Inducement for adding pectin or acid in quantities greater than required to supply the natural deficiency of the fruit juice is eliminated by fixing the minimum fruit juice content. The name of the fruit or fruits present as well as spices, chemical preservatives, honey, or corn syrup used must be indicated on the label.

Standards for jams and preserves are similar to those for jelly except that fruits are used rather than fruit juice ingredients, and mint flavor and green coloring are not optional ingredients. The fruit mixture is concentrated by heat to such an extent that the total soluble solids content is not less than 65% for certain specified fruits, and 68% for others.

It is extremely difficult to account for the behavior of gels formation. Gels are a form of matter intermediate between a solid and a liquid. They consist of polymeric molecules cross-linked to form a tangled, interconnected molecular network immersed in a liquid medium. The water, as a solvent, influences the nature and magnitude of the intermolecular forces that maintain the integrity of the polymer network; the polymer network holds the water, preventing it from flowing away in acid medium; pectin with sugar affects the pectin water equilibriumand formsanetwork offibers throughout the jelly. This structure is capable of supporting liquids.

TYPES AND VARIETIES OF FRUITS

A jam manufacturer can choose a fruit from among the following five categories:

1. Fresh fruit
2. Frozen, chilled, or cold stored fruit
3. Fruit or fruit pulp preserved by heat
4. Sulfited fruit or fruit pulp, i.e., fruit preserved with sulfur dioxide
5. Dried dehydrated fruits

1Fresh fruits generally give the best jams. As pectin is the main ingredient in the fruit that gives a set to the jam, it is preferable to use a slightly underripe fruit that is rich in pectin along with the ripe fruit to secure the desirable gelling effect in the jam. Apple, pear, sapota, apricot, loquat, peach, papaya, karonda, plum, strawberry, mango, tomato, grapes, and muskmelon have been used for preparation of jams. It can be prepared from a single fruit or a mixture of two or more.

In jelly making, pectin is the most essential constituent. Although there is difference of opinion about the exact nature of pectin, it is generally accepted that pectin forms jellies, when mixed and boiled with proper amounts of sugar, acid, and water. All these constituents must be present in a particular proportion for making a good jelly.

Guava, sour apple, plum, grapes, karonda, wood apple, loquat, papaya, and gooseberry are generally used for preparation of jelly. Apricot, pineapple, strawberry, raspberry, etc. can be used but only after addition of pectin powder, because these fruits have low pectin content. Fruits can be divided into four groups on the basis of their pectin and acid contents.

1. Richin pectin and acid: sour and crab apple, grape, sour guava, lemon, orange (sour), plum (sour), jamun.
2. Rich in pectin but low in acid: apple (low acid varieties), unripe banana, sour cherry, fig (unripe), pear, ripe guava, peel or orange, and grapefruit.
3. Low in pectin but rich in acid: apricot (sour), sweet cherry, sour peach, pineapple, and strawberry.
4. Low in pectin and acid: ripe apricot, peach (ripe), pomegranate, raspberry, and any other over-ripe fruit.

Types and varieties of fruits selected should be used without undue delay for making jam and jelly because, if kept for a long time, degradation of pectin proceeds rapidly.

Peels

Apart from the sorting and removal of leaves, stalks, and undesirable portions of the fruit, which can only be done by hand, most fruits require treatment of some kind before they enter the boiling pan. For example, strawberries require sometime light crushing, plums require heating with minimum of water until soft. When it is required, cherries are similarly treated. Currants are passed through machines that remove the stalks, gooseberries are whirled in machine.

Sour and bitter oranges are utilized by hand or machine and the peel cut of any desirable size or shape by special machines. Sometimes admixture of certain proportion of grapefruit or lemons is used, peel of which constitutes only a quarter to a half if utilized in proportion to the total weight of fruit. These slices need to be softened either by prolonged boiling, or more rapidly by heating. The slices may be covered with water and cooked until tender, the water being changed at least twice during cooking. It can be done in autoclaves under pressure of 1 atm. More rapid softening is attained by boiling the peel in a solution of carbonate of soda or ammonium hydroxide, which removes from the cell wall certain substances that otherwise render the peel tough after boiling with sugar. Ammonia is less drastic in its action than that of soda and is preferable for use since less danger of the peel breaking up into small fragments, which is due to the fact that soda dissolves both hemicelluloses and pectic subtances, whereas the former are said to be insoluble in ammonia. It would appear that some more research is required as to the precise causes of the softening.

The time required to soften peel in an autoclave depends on the size of the slices and on the pressure and temperature employed. Some discoloration may occur if too high pressure is used. To avoid handling the peel after it has been softened, Morris suggested placing the peel intended for softening in perforated baskets of non-corrosible metal, each containing sufficient peel for one batch. After having been cooked in ammonium hydroxide or soda, the peels may be re-cooked for a short time with a weak solution of citric acid to remove any traces of the hydroxide.

Gelling Agents

Gelling agents are used in the food industry in a wide range of products both traditional and novel, and this use is increasing rapidly with the increase of convenience foods. An ideal gelling agent should not interfere with the odor, flavor, or taste of the product to which it is added. Improvements to existing and development of new ones require basic understanding of the processes of gelatin and the properties of gels at the molecular level.

Gels are a form of matter intermediate between a solid and a liquid. They consist of polymeric molecules cross-linked to form a tangled, interconnected molecular network immersed in a liquid medium. The polymer network holds the water, preventing it from flowing away. In gels, the molecules are held togetherbyacombinationofweak intermolecular forces such as hydrogen bonds, electrostatic forces, Vanderwaals forces, and hydropho-bic interactions. The cross-linkages are not point interactions but involve extensive segments from two or more polymer molecules, usually in well-defined structures called junction zones. The gelation process is essentially the formation of these junction zones.

The physical characterizations of gel are the consequence of the formation of a continuous three-dimensional network of cross-linked polymer molecules on a molecular level; an aqueous gel consists of three elements:

1. Junction zones where polymer molecules are joined together.
2. Interjunction segments of polymers those are relatively mobile.
3. Water entrapped in the polymer network.

Gels are always formed in an aqueous environment. Thus, the interactions of protein and polysaccharides with water are by themselves important factors in the gelation process. Both types of polymers are strongly hydrated in aqueous solution, so that some water molecules are so tightly bound that they fail to freeze even at temperatures as low as -60%C. Although the formation of a stable intermolecular junction is a critical requirement for gelation, some limitation on the interchain association is also necessary to give a hydrated network rather than an insoluble precipitate.

It is important to know the condition forthe onset of gelation in technological processes involving gelling food products. Several methods are used to characterize this change in consistency. Physically, the critical stage of gelation may be monitored from the loss of fluidity or from the rise of the elastic property of the growing network.

Gelatin is a water-soluble protein formed by initial degradation of collagen from animal skin and bones. Gelatin jellies have a rather soft or rubbery texture. For these, it is normal to use an additional gelling agent such as thin boiling starch. This involves the texture incidentally. Gelatin gels forms reversibly on cooling a gelatin solution. It is now well established that the protein molecules are cross-linked to form a network by junction zones, where the protein chain have partly refolded in the collagen triple helix structure.

Agar/alginates are the major structural polysaccharides of algae. Agar jellies have a very soft texture. Straight agar jellies have a characteristic "shortness" that may be modified by the addition of gelatin, gum Arabic, pectin, starch, etc. Alginates with a high ratio of poly-β-D-mannuronic acid (M) and poly-α-L-guluronic acid (G) form weak, forbid gels, whereas low M/G alginates give transparent, stiff, brittle gels, and the gel strength depends on the nature of the divalent cation.

Gum Arabic is the most water-soluble of the natural gums (up to 50%) and their solutions are of relatively low viscosity. Other advantages of gum Arabic are its absence of odor, color, and taste. Hard jellies can be produced with gum Arabic.

Unmodified starches, produced by wet milling of field corn, supply the major amount of thickening material. Modified starch is starch that has undergone one of the varieties of treatment to alter its physical property and/ or functionality. They are used to extend the bodying or gelling effect of normal starches, to modify gelling tendencies, and to improve texture. Starch is an essentially linear polymer of α-(1→4) linked D-glucose. Starch gels consequently have a composite structure of open, porous amylopectin molecules threaded by an amylose matrix. Thus, actual gel-forming polymer in starch is amylose. The molecular weight distribution of amylose depends on the plant source and molecular weights of several millions with broad distribution have been reported.

Pectin is the most frequently used hydrocolloids in processed fruits. Jams and jellies are the major food type using larger amounts of pectin. Pectin is a class of complex hetero polysaccharides found in the cell walls of higher plants, where they function as a hydrating agent and cementing material for the cellulosic network.

When pectin-rich plant materials are heated with acidified water, the protopectin is liberated and is hydrolyzed into pectin that is readily soluble in water. It happens in plant tissues during ripening of fruits with the aid of an enzyme protopectinase. As the ripening of fruit proceeds, more and more of the insoluble protopectin is converted into soluble pectin. Their composition varies with the source and conditions of extraction, location, and other environment factors. Pectic substances in the primary cell wall have a relatively higher proportion of oligosaccharides chain on their backbone, and the side chains are much longer than those of the pectin of the middle lamella.

Pectins are primarily a polymer of D-galacturonic acid (homopolymer of [1→4] a-D-galacto pyra-nosyluronic acid units with varying degrees of carboxyl groups methylated estrified) and rhamno-galacturonan (hetero polymer of repeating [1 →2] α-L-rhamnosyl [1-L] →D-galactosyluronic acid dis-accharide units), making it an →D-galacturonan. The molecule is formed by L-1,4-glycosidic linkages between the pyranose rings of D-galacturonic acid units. As both hydroxyl groups of D-galacturonic acid at carbon atoms 1 and 4 are on the axial position, the polymer formed is 1,4-polysaccharide.

Pectic acid is composed mostly of colloidal poly-galacturonic acid molecules, and is essentially free from methyl ester groups. The salts of pectic acids are either normal or acid pectates, whereas pectinic acid ones are colloidal polygalacturonic acids containing more than a negligible portion of methyl ester groups. Pectinic acid under suitable conditions is capable of forming gels with sugar and acid or, if suitably low in methoxyl content, with certain metallic ions. The salts of pectinic acids are either normal or acid pectinates.

Studies on esterified residue in pectin claimed that they are randomly distributed or non-randomly distributed. However, ion exchange chromatography showed a random distribution of change in citrus pectin that had undergone acid catalyzed deesterification. Such disparate findings may, impart, be due to the length of galactur-onate residues being examined or due to differences in pectin source.

Polygalactoronic acid could be considered as a rod in solution, whereas pectins are segmented rods with flexibility at the rhamnose tees. The size, charge density, charge distribution, and degree of substitution of pectin molecules can be changed biologically or chemically.

The most unique and outstanding property of pectin is their ability to form gel in the presence of Ca^2+ ions in sugar and acid solution. Degree of esterification, attached chain of neutral sugars, acety-lation, and cross-linking of pectin molecules also affect the texture of pectin gels. Pectin chains carry negative charge and the change density is higher at higher pH and lower degree of methoxylation (DM). Conformation of the pectin molecules is not affected by the branching, but side branching in pectin can result in significant entanglement in concentrated solutions. Depending on the DM, pectins are classified into (1) low methoxy (LM) (25–50%) and (2) high methoxy (HM) (50–80%) pectins, and form gels of two types with occasional intermediates. They are called acid and calcium gels and formed from HM and LM pectins, respectively. In HM pectins, the effect of sugars depends specially upon molecular geometry of the sugar and the interactions with neighboring water molecules. Non-covalent forces (i.e., hydrogen bonding and hy-drophobic interactions) are believed to be responsible for gel formation in HM pectins. In LM pectins, gel is formed in the presence of Ca^{2+}, which acts as a bridge between pairs of carboxyl groups of pectin molecules. The two kinds of pectins are relatively stable at the low pH levels existing in jams and jellies. HM pectin are used to form gels in acid media of high sugar content and LM pectins are used in products of lower sugar content. Pectins can be further divided into rapid-set, medium-set, and low-set pectin, depending upon the time the gel takes to set.

The functionality of pectin molecules is determined by a number of factors, including DM and molecular size. Pectins of100– 500 grades are available in the market. Their application as a food hydrocolloid is mainly based on their gelling properties. Selection of pectin for a particular food depends on many factors, including the texture required, pH, processing temperature, presence of ions, proteins, and the expected shelf life of the product. Combinations of gelling agents are often used because a combination gives a desirable texture. For this reason, mixed systems are of great technological importance. The physical chemistry ofmixed system is obviously very complex. However, there has recently been significant progress in developing a suitable theoretical framework, at least for two component systems.

Sweetening Agents

Sweeteners are used in fruit processing for many functional reasons as well as to impart sweetness. They add flavor, body, bulk, and control viscosity that contribute to texture and prevent spoilage. It binds moisture in fruits that is required by detrimental microorganism. Too little sugar prevents gelling and may allow yeast and mold growth. Sugar serves as a preserving agent and aids in jelling.

Sucrose, commonly known as table sugar (or refined sugar), is the standard against which all sweeteners are measured in terms of quality, taste, and taste profile. However, glucose syrups have been widely used as a part of sugar source in recent years. An invert sugar component is necessary to prevent sucrose crystallization in high solids jellies and jams during storage. Such crystallization is rare in products containing less than 68% solids. If the concentration of sugar is high, the jelly retains less water resulting in a stiff jelly, probably because of dehydration.

The optimum percentageofinvert sugarisbetween 35% and 40% of the total sugar in the jam. During the process of inversion, molecular water is taken into the sugars; that is the reason why 95 parts of sucrose yield 100 parts of invert sugar. The rate of inversion is influenced by three factors:

1. Hydrogen ion concentration (pH) of the product
2. Boiling temperature
3. Boiling time

During the process of boiling, sucrose undergoes a chemical change. Cane and beet sugars are non-reducing sugars. However, when boiled with acid or treated with some enzymes, sucrose is converted into two reducing sugars, namely dextrose and levulose in equal parts. Sucrose has a molecular weight of 342, invert sugar 360, the difference of 18 being the molecular weight of water molecule added during inversion.

Invert sugar syrup canbeproduced from sucrose by the action of an acid. After the process is completed, a suitable amount of sodium bicarbonate is added in order to neutralize the acid, as it is frequently undesirable to use strong acid in a food containing sugar. Methods using hydrochloric, tartaric, or citric acid are satisfactory when pure white sugar is used. Lower grade of sugar may require larger amount of acid. A chemist should be consulted for the quantities of acid needed.

The addition of sugar in LM pectin gels can increase gel strength and reduce syneresis. For HM pectin gels, a number of other sugars, alcohols, and polyols will permit gelation. From a practical point of view, it may be advantageous to substitute other sugars for sucrose, either because of cost, or to reduce the likelihood of crystallization, or for flavor modification. Partial replacement of sucrose with other sugars such as maltose, glucose, syrups, or high fructose corn syrup altered the setting times and certain rheological properties of model gels. For example, addition of maltose reduced the setting time and extended the pH range of gelation, while fructose delayed setting time. Partial or complete replacement of sucrose with other sugars alters the water activity of the system and can modify the hydrophobic interactions contributing to gelation.

HM pectin gel forms only under acidic conditions and when the sugar content is at least 55%. Low pH suppresses dissociation of free carboxylic acid groups, reducing their electrostatic repulsion, while sugars stabilize hydrophobic interactions between the methyl ester group. Junction zone size and the standard free energy of gelation increase as the degree of esterification increases, being proportional to the square of the degree of esterification. As pectin ester content is lowered to 50%, jelly strength increases, but only at progressively lower pH values.

Esterification levels of pectins can also have an effect on the flavor perception of the jellies. For a substance to be tasted, it must contact the taste buds. Therefore, if a gel delays diffusion of the substance to the taste bud surface perception may be controlled by diffusion and not by the taste reaction. Guichard et al. found, at similar gel consistencies, high methoxyl pectin reducestaste intensity more than low methoxyl ones.

Acidulants

Acidulants are acids that either occur naturally in fruits and vegetables or are used as additives in food processing. Acidulants serve many functions in fruit processing. They perform a variety of functions in fruit processing, the primary being acidifier, pH regulator, preservative and curing agent, flavoring agent, chelating agent, buffering agent, gelling/coagulating agent, and antioxidant synergist.

The choice of the particular acid for any specific food application is dependent upon a number of factors. Since each acid has its own unique combination of physical and chemical properties, the choice should be made on the basis of the qualities required. The simplest possible use of a food acid involves decreasing the pH only. As an example, it might be desired simply to acidify a solution where pH adjustments are all that are required. In many cases, the selection is based upon the ability of the acid to bring out and enhance the food flavor. For such purposes, compatibility and blending are important factors in the choice of acid. Each acid behaves somewhat independently of pH and not necessarily preferred for another. Thus, citric acid is preferred in most fruits, tartaric acid for grape, and malic acid for apple. Since taste stability cannot always be predicted, the product should be examined both before and after storage, when blending or substituting the acids used.

Apples differintheir organic acid profiles. Frauzen and Helwert found oxalic 0.001, succinic 4.22, malic 69.57, citric 24.95, and lactic 1.18% in apples. Nelson, on the other hand, found a much less complicated mixture in American apples, the acid being entirely malic. The total acidity may be about 1.2% (as malic acid) in apples. It shows that the same kinds of fruits may vary, not only in the amount but also in the kind of acid they contain. The H ion concentration is affected very considerably by buffering substances such as organic bases and inorganic and organic salts and is of the utmost importance in connection with fruit preservation and a determining factor in the formation of the pectin–sugar–acid gel.

It has been found that gel formation occurs only within a certain range of hydrogen ion concentration, the optimum acidity figure for ions and jellies being around pH 3.0. The gel strength fouls slowly on decreasing and rapidly on increasing thepH value. Beyond pH 4, no jelly formation occurs in the usual soluble solid range. The pH value is also critical in determining the temperature at which jellies set. Insufficient acidity is one ofthe common causes ofjelly failure. The pH value of jelly should be taken when the jelly is concentrated sufficiently to pour. Different juices will require different amounts of additional acid depending upon the original acidity of the juice and buffering capacity of the juice. The pH may be adjustedtoattain optimumflavortocontrolormodify the rate of setting and to modify the degree of sugar inversion.

Control of pH is critical to successful gel formation with pectin, particularly in the case of HM pectins. Low pH increases the percentage of unionized car-boxyl groups, thus reducing electrostatic repulsion between adjacent pectin chains. Rapid-set pectins with high degree esterification will gel at higher pHs than the one with lower degree esterifications, slow-set pectin; however, this difference is slight, with the optimum pH for slow-set pectins being about 3.1 and for rapid-set pectins being 3.4. Substitution of other sugars for sucrose, by modifying hydrophobic interactions between chains, allows gels to be formed at somewhat higher pHs. Since they relyon calcium bonding to effect gelation, low methoxyl pectin can form gels at higher pHs than high methoxyl pectins. Gels can be made atpHvalues near neutralite, which is advantageous in producing dairy-based products.

Coloring and Flavoring Agents

The color of jam is a factor of considerable importance.Agood jam should appeal to the eye as well as to the palate. No coloring is required for jams produced from fresh fruit, provided the boiling time is short and the heat not excessive. The natural color of the fruit is, however, always affectedbypresentation with SO_2 andinsome cases by the process of boiling, necessitating the addition of artificial color. The aim should be to restore the original natural appearance. Only permitted edible food colors should be used.

Coal tar colors are most frequently used to a lesser extent. It is essential that the colors should be intensive, readily soluble, and stand up to high concentration in solution. As a rule, acid colors are of higher stability than basic ones. Color should be acid proof; many are affected by acids and particularly by sulfur dioxide present in glucose and fruit pulp. They must also be light proof and have certain stability to heat. All colors suffer from prolonged and excessive heat and should therefore be added at the last stage of the boiling process.

Traditionally, coal tar dyes have been used to color jams and marmalades, particularly those prepared from sulfited fruit. Color testification of jams with natural colors, notably with anthocyanins from grape skins, is increasing as artificial additives become less acceptable to consumers. Colors available in powder form should be prepared just before the time of addition, since many colors have a tendency to precipitate on standing. When dissolving, the color is mixed to a paste with a little cold water then adding the required amount of boiling water and stirring well.

Ordinarily jams do not require the addition of flavors. If desired, they may be added when jam boiling is nearing completion. Citrus oils and fruit volatile compounds (recovered in the concentration of fruit juices) can be added toward the end of the process to improve product aroma.

Product Type and Recipes

According to the U.S. federal standard, a jam should have a minimum of 45 parts of prepared fruit and 55 parts of sugar, concentrated to 65% or higher solids, resulting in a semi-solid product. Jellies are similar to jams with 45 parts of clarified fruit juice and 55 parts of sugar, resulting in a minimum of 65% solids. Both categories can utilize a maximum of 25% corn syrup as sweetener as well as pectin and acid to achieve the gelling texture required.

On the basis of specification and the allowance for soluble solids such as acids and pectin, it will be seen that the finished jam or jelly, no matter what the grade, will contain approximately the quantity of sugar necessary to give the maximum strength to the pectin–sugar–acid gel. It may be assured that about 3–5% of the total weight of the jam is represented by sugar derived from the fruit; hence, about 65% is added sugar. This figure will therefore be based on the nature of the jam or jelly that the manufacturer wishes to produce. If the fruit that is being used is deficient in pectin or acid or both, or if the quantity of fruit in the recipe is less, it may be necessary to add pectin or acid, so that a gel with sufficient firmness can be obtained.

Some typical recipes found useful in large-scale manufactureofjams mentionedbyGiridhari Laletal. are given below.

To prepare jam, the requirement of pulp (fresh or canned) is 75 kg, with sugar 75 kg, citric acid 35 g, pectin 150 grade 565 g, and pineapple essence 75 ml.

To prepare orange jam, 50 kg of lye peeled segments require 50 kg of sugar, citric acid 250 g, pectin 150 grade 375 g, and sweet orange essence 50 ml.

To prepare mango jam, 40 kg of mango pulp requires sugar 40 kg, pectin 150 grade 500 g, citric acid400 g, and mango essence 70 ml.

To prepare apple jam, 40 kg of apple pulp requires sugar 44 kg, pectin 150 grade 400 g, citric acid 500 g, and apple essence 60 ml.

To prepare mixed fruit jam, blends of mango, pineapple, orange, apricots, papaya, guava, etc., and equal weight of sugar to that of blended pulp taken, and citric acid to the extent of about 0.75–1.0% by weight of blended pulp containing pectin to the extent of 0.5–1.0% by weight of blended pulp are required depending upon the fruits used. A blend of predominantly red food grade colors may be added along with an appropriate essence to the desired extent.

Jams from cherry, mulberry, strawberry, muskmelon, jack fruit, cashew apple, etc. also can be made in the usual way. It may, however, be necessary to vary slightly the fruit sugar ratio and the percentage of acid added. In some cases, it may be useful to supplement the flavor of the jam by adding extra fruit flavor.

To prepare jelly from guava, ripe fruits are washed well and cut into thin slices and covered with an equal weight of water containing 2 g of citric acid for every kg of guava taken. The mixture is heated gently for about 30 min to extract pectin from the slices. The extract is drained and to the residue, water equal to one-fourth of the weight of slices is added and a second extract is taken. The two extracts are combined and strained through a piece of thick cloth to remove the coarse particles. The strained extract is allowed to settle overnight in a tall vessel. The clear supernatant liquid is tested for its pectin content. The required quantity of sugar (generally an equal quantity) is added to the extract and the mixture boiled down to the desired consistency. The finished jelly is packed in glass jars or in cans. By adopting the procedures described for guava jelly, jellies can be prepared from fruits such as apple, grape, jamun, jack fruit, etc.

Uncooked jam and jelly recipes areanewer method of preserving fruit produce to keep fresh from the wine flavor. These are also called freezer jams, since they need to be stored in the freezer to preserve them. In these recipes, the uncooked fruit is just mixed with sugar, lemon juice pectin, and stirred. Powdered pectin is usually dissolved in water first and then added to the fruit. Care is taken to stir well to dissolve as much sugar as possible, so that it would not have a sugary or gritty jam or jelly. The jam is then usually poured into covered plastic freezer container or can or freezer jars and allowed to set at room temperature. It may take 24 h for the product to jell properly. Recipes for uncooked freezer jams and jellies come with all the pectin products available and the methods of preparation vary slightly for each.

Method of Manufacture

Jam and jelly making is essentially a process that lends itself to be controlled by a chemist. The basic important steps in selection of fruit, boiling for setting, filling of the finished product, and packaging will be discussed below. Manufacture of jams and jellies may be considered rather simple; however, unless scientific approaches are not adhered to, the finished product will not be perfect.

Fruit Preparation

Fruits for jam making should be fully mature, possess a rich flavor and be of the most desirable texture. Fruits are washed thoroughly with water to remove any adhering dirt.Ifthe fruit has been sprayed with lead or arsenical sprays, it should be washed in a warm solution of 1% hydrochloric acid and then rinsed in water.

All berries must be carefully sorted and washed. Strawberries must be stemmed; peaches, pears, apples, and other fruits with heavy skin must be peeled, while apricots, plums, and fresh prunes can be pitted by machine. Stone fruits such as plum and apricots require a very heavy pulping screen because of abrasive action of the pits. Berries should not be softened by boiling before the addition of sugar, but need only to be crushed.

The proportion of sugar to fruit varies with the variety of the fruit, its ripeness, and the effect desired, although the most common ratio of sugar to fruit is 1:1. This is usually a suitable ratio for berries, currants, plums, apricots, pineapple, and other fruits. Sweet fruits with low acidity such as ripe peaches, sweet prunes, and sweet varieties of grapes normally require less than an equal weight of sugar.

Jams may be made from practically all varieties of fruits. Various combinations of different varieties of fruits can often be made to advantage, pineapple being one of the best for blending purposes because of its pronounced flavor and acidity.

Fruits for Jelly

Fruits required for jelly making should contain sufficient acid and pectin to yield good jelly without the addition of these substances, although often in commercial practice, this ideal condition is not attained. Some fruits contain enough of both pectin and acid for the purpose, while others are deficient in either one or both the substances.

In the case of fruits deficient in pectin but rich in acid, fruits rich in either pectin or commercial pectin should be added. Both these methods have their own merits. Combining of fruits rich in acids with those rich in pectin is expensive than using acid or commercial pectin to supplement the deficiency, but the drawback is that the flavor of the jelly is affected. Special care is, therefore, necessary to ensure that the fruits are mixed in proper and judicious proportions; on the other hand, commercial pectin does not have any deleterious effect on the final flavor of the jelly made from any particular fruit.

Of the fruits rich in pectin and acid, table varieties of apples, sour blackberries, currants, lemons, limes, grapefruits, sour oranges, sour guavas, and sour varieties of grapes, plums, and cherries are good examples. Of the fruits low in acid but rich in pectin are sweet varieties of cherries, unripe figs, ripe melon, and unripe bananas. Fruits that are rich in acid but low in pectin are apricots and most varieties of strawberries. Fruits that may be classified as containing a moderate concentration of both acid and pectin are ripe varieties of grapes, ripe blackberries, ripe apples, and loquats. Fruits low in both acid and pectin are represented by pomegranates, ripe peaches, ripe figs, and ripe pears.

It is customary to blend fruits deficient in acid or pectin, or both, with fruits that have an abundance of the required constituents.

Boiling

Boiling is one of the most important steps in the jelly making process, as it dissolves the sugar and causes union of the sugar, acid, and pectin to form a jelly. It usually causes a coagulation of certain organic compounds that can be skimmed from the surface during boiling, and their removal renders the jelly clearer. The principal purpose of boiling is to increase the concentration of the sugar to the point where jelling occurs.

The boiling operation, while normally being a necessary step in jelly making, should be as short as possible. Prolonged boiling results in loss of flavor, change in color, and hydrolysis of the pectin; consequently, it is a frequent cause of jelly failure. In jam making, the fruits resistant to boiling are desirable to concentrate the product by evaporation of excess moisture. Boiling in commercial practice is usually conducted in open steam jacketed stainless steel kettles. During boiling, the juice or pulp should be skimmed, if necessary, to remove coagulated material and should be stirred to cause thorough mixing. The boiling is continued until on cooling, the product will form a jam or jelly of the desired consistency. The concentration of the mixture when this point is reached will depend upon several factors viz., the concentration of pectin, the concentration of acid, the ratio of sugar to pectin and acid, and the texture desired.

The most common method of determining the end point is by allowing the liquidto sheet from awooden paddleor a large spoon.Ifitdrips from the instrument as thin syrup, the process is not complete. If it partly solidifies and breaks from the spoon in sheets, or forms jelly-like sheets on the side of the spoon, the boiling is considered to be complete.

Boiling of the sugar solution in the presence of the fruit acid results in inversion of some of the cane sugar. Hence, a jelly that is boiled for a long period is less liable to develop crystals of sucrose than a jelly boiled for a short time. Prolonged boiling may result in loss of flavor through evaporation, hydrolysis, or other forms of decomposition.

Sometimes jam is boiled in a vacuum pan at a lower temperature in the range of 65–76%C. The vacuum boiling or cooking minimizes the undesirable changesincolor and prevents lossofvitaminC.How-ever, the jam mixture has to be boiled for a long time to soften the fruit pieces resulting in some loss of flavor which can, however, be restored by recovering the volatile esters and putting them back into the jam. Proper control of boiling is necessary to avoid over concentration of soluble solids, over inversion of sugar, and hydrolysis of pectin. Manufacture of jam can be considered rather a simple process, but unless controlled properly may lead to undesirable results.

Filling

The finishing of jams comprises three main steps: (1) pre-cooling prior to filling, (2) filling, and (3) cooling after filling.

Prolonged heating affects the appearance as well as the keeping quality of the finished product. As the inversion of sugar is greatly influenced by the temperature, it is obvious that an efficient cooling system is necessary to control and check the process. Difficulties are also experienced in filling, as some fruit varieties show a tendency to float, the most susceptible being strawberry, cherry, black currant, and stone fruit jams.These jams should be cooled until they are near setting point, but great care must be taken not to exceed the limit, otherwise the set will break and the jam curdles, more so in the case of jellies.

The efficient method in trade is filling on a roller conveyor. The empty jars are packed into trays, each one holding a certain number. The trays are then moved on roller conveyer to the filling operators. The trays containing filled jars are moved to cooling room after the preliminary setting has taken place. This prevents discoloration due to caramalization of the sugar.

After being filled, jam in jars must not be cooled too quickly. As far as canned jam is concerned, the procedure is simple enough, the cans being passed through a water bath. Glass jars and large containers have to be cooled by air. This can be done by passing them slowly through atunnelfitted withanair blast or keeping them in a cooling room constructed on the same principle until the jam is well set. Automatic filling machines that measure a definite volume of jelly into each container are in general used in large factories. They greatly reduce the cost of filling as compared to that of hand filling and give more uniform net contents.

Packaging Materials

The final stage of the production process is packaging. A wide variety of sizes and shapes of containers are used for jellies. Glass is the usual material, although enamel-lined tin cans and special containers are also used.

Jelly should be hermetically sealed in glass containers. A paraffin seal is not adequate to prevent spoilage of the product. Container filled scalding hot (in excess of 83%C) need not be pasteurized, as the hot-filled jelly itself will sterilize the container. The jar should be filled to at least 90% full leaving not more than 1.25 cm head space. The scalded lids should be loosely placed on the containers immediately following filling, and then tightened firmly within 2–3 min. This allows time for exhausting of air from the head space. The steam in the head space condenses when the jelly cools, creating a vacuum seal. Capping with superheated steam injection is often used to attain a hermetic seal. Where the product is not filled sufficiently hot to ensure head space sterilization, or where superheated steam injection is not used, a post-capping sterilization treatment is employed.

Some jellies form during boiling and filling a layer of bubbles on the surface of the jar of hot jelly. The jelly should be quickly skimmed while in the kettle just prior to pouring. If the jelly can be drawn from the bottom of the kettle, clear jelly can be filled into the jars.

If jelly is to be poured into glass jars, the sides of the jar should be smooth so that the jelly can be turned out without breaking its shape or structure. Before pouring the jelly, glass containers should be warmed to prevent breakage. After filling, the jars should be cooled rapidly to about 21%C. Pectin jellies set more quickly at this temperature than at lower temperatures. If the jelly fails to set or is weak, it is placed in a drier to evaporate the excess of water in it and promote setting.

Deterioration in storage is now largely prevented by hermetically sealing the jars, while still hot, in sterile manner using metal caps fitted with rubber gaskets. Many patented caps of this kind have been devised which can be placed in position, sterilized by steam jets in a steam box and, finally, held firmly by creating a partial vacuum in the head space of the jars, either in a specially constructed vacuumizing chamber or merely by screwing them up tightly while still hot. Jars sterilized and sealed in this way form an ideally hygienic package and are to a very large extent independent of storage conditions.

Two types of vacuum-sealed jars are in common use. In one of these, the seal between the glass and jar cover is made with a rubber composition ring attached to the cap. This composition melts during pasteurization, and after cooling of the jar and contents, it solidifies to form an air-tight seal. The lid must be held in place with a clamp during pasteurization. The second type of jar is sealed with a rubber gasket similar to a fruit jar rubber, but this rubber is pressed against the side of the jar rather than the top. It is held in place by friction and the cap is rolled in much the same manner as an ordinary sanitary can top. The cap needs no clamp to hold it in place during pasteurization. A similar lid is pressed into position but is not rolled.

Small enamel-linked jam cans and gallon cans are sometimes used for jelly. Wooden tubes or buckets are often used for cheap jellies for baker's use, the product usually being preserved with sodium benzoate. The inside of cans used for jellies made from fruits of red color should be heavily lacquered to prevent bleaching of the color by tin salts. Some of the jams are sold in cans. Glass containers are almost universally used for jams and jellies. For jams also, the processes of filling and sealing are done by automatic machinery as described for jellies.

Future Research Needs

Jam and jelly production relies on the native pectins of incorporated fruit for gel formation. Modern manufacturing requirements of uniform gel strength and appearance preclude reliance on fruit maturity and variety. In spite of the current availability of other gelling agents, pectin remains the universal choice for jams and jellies, in part because of its presence as a natural fruit ingredient, and also because of the characteristic consistency that pectin imparts to a gel. The joint FAO/WHO committee on food additives recommended pectin as a safe additive with no limit on acceptable daily intake, except asdictated by good manufacturing practice.

In recent years, pectin has been used as a fat or sugar replacer in low calorie foods. It is estimated that 80–90% of commercial pectin production which totals 6–7 million kg is used in the production of jellies and jams. In spite of its availability in a large number of plant species, commercial sources of pectin are very limited. There is therefore a need to explore other sources of pectin or modify the existing sources to obtain pectin of desired quality attributes. Modern food sciences such as genetic engineering can be used to modify pectin *in vivo*. Current knowledge of the molecular basis of gelation in pectin has helped us to understand some aspects of this complex phenomenon. There are still some areas where our knowledge is scanty. So a systematic study of these observations will help in understanding the reaction processes in pectin gel formation, resulting in better control of processes and products.

References

Abers, J.E., Wrolstad, R.E. 1979. Causative factors of colour deterioration in strawberry preserves during processing and storage. *Journal of Food Science*. 44, 75–78.

Anon. 1983. Jams and jellies. *Institutional Distribution*. 19(12), 218, 222–224.

Baker, R.A., Berry, N., Hup, Y.H. 1996. *Fruit preserves and jams in processing fruits—Science and Technology,* Vol 1. Biology, Principles and Application. Technomic Publishing Co., Inc.

Giridhari Lal, Siddappa, G.S., Tandon, G.L. 1986. *Preservation of fruits and vegetables*. Indian Council of Agricultural Research. New Delhi.

Kratz. 1993. Jam, jellies, marmalade II. The phenomenon of syneresis and method of manufacturing. *Food Marketing & Technology* 17–21.

7

Dehydration and Concentration Methods

The preservation of foods through the removal of water is probably one of the oldest techniques. In the past, the process was simplified by directly exposing the product to sunlight. The crop was spread on the ground directly or over sacks or mats made from plant leaves. Today, the quality of dried products has improved thanks to a number of factors, including the following.

— The use of dehydrating equipment for solar and artificial drying, which increases the efficiency of dehydration.

— The use of chemical pre-treatment to better preserve the colour, aroma and flavour of the products.

The fundamental principle on which dehydration is based is that at low moisture levels, the water activity drops down to levels at which neither micro-organisms nor deteriorating chemical reactions can develop.

Generally speaking, vegetables with less than 8% moisture and fruits with less than 18% residual moisture are not favourable substrates for the development of fungi, bacteria or important chemical or biochemical reactions. There are reactions, such as non-enzymatic browning, which may develop at slower rates in low water-level environments, but that requires high temperatures. Other reactions include fat oxidation, which may occur at very low water levels, but which is accelerated by light and temperature. The container and environment in which the dehydrated products are kept are thus extremely important to guarantee good preservation.

It is very important to avoid contamination by dust and other substances that may be the carriers of micro-organisms which are resistant to low moisture levels, as for example the excrements or urine of rodents or domestic animals, chemical products, pesticides and others. The sites used for the drying process must also be very carefully chosen.

The drying time and the product's final moisture level will depend on the location of the dryer, the climatic conditions of the place and the characteristics of the product. Material cut in small sections and with a greater drying surface will dry more rapidly.

The drying process must be handled with great care, if one wishes to obtain a quality product. Often, the drying must take place in the shade in order to preserve the product's sensory qualities, like colour, aroma and texture.

Water and Water Activity (A_w) in Foods

Micro-organisms in a healthy growing state may contain in excess of 80% water. They get this water from the food in which they grow. If the water is removed from the food it also will transfer out of the bacterial cell and multiplication will stop. Partial drying will be less effective than total drying, though for some micro-organisms partial drying may be quite sufficient to arrest bacterial growth and multiplication.

Bacteria and yeasts generally require more moisture than moulds, and so moulds often will be found growing on semi-dry foods where bacteria and yeasts find conditions unfavourable; example are moulds growing on partially dried fruits.

Slight differences in relative humidity in the environment in which the food is kept or in the food package can make great differences in the rate of micro-organism multiplication. Since micro-organisms can live in one part of a food that may differ in moisture and other physical and chemical conditions from the food just millimetres away, we must be concerned with conditions in the "microenvironment" of the micro-organisms. Thus it is common to refer to water conditions in terms of specific activity.

The term "water activity" is related to relative humidity. Relative humidity is defined as the ratio of the partial pressure of water vapour in the air to the vapour pressure of pure water at the same temperature. Relative humidity refers to the atmosphere surrounding a material or solution.

Water activity or aw is a property of solutions and is the ratio of vapour pressure of the solution compared with the vapour pressure of pure water at the same temperature. Under equilibrium conditions water activity equals:

$$a_w = RH / 100$$

When we speak of moisture requirements of micro-organisms we really mean water activity in their immediate environment, whether this be in solution, in a particle of food or at a surface in contact with the atmosphere.

At the usual temperatures permitting microbial growth, most bacteria require a water activity in the range of about 0.90 to 1.00.

Some yeasts and moulds grow slowly at a water activity down to as low as about 0.65.

Qualitatively, water activity is a measure of unbound, free water in a system available to support biological and chemical reactions. Water activity, not absolute water content, is what bacteria, enzymes and chemical reactants encounter and are affected by at the micro-environmental level in food materials.

Two foods with the same water content can have very different a_w values depending upon the degree to which water is free or otherwise bound to food constituents.

Thus, this food, at the temperature for which this absorption isotherm was established, will ultimately attain a moisture content of 20% at 75% RH (relative humidity). If this food was previously dehydrated to below 20% moisture and placed in an atmosphere of 75% RH, it would absorb moisture until it reached 20%. Conversely, if this food was moistened to greater than 20% water and then placed at 75% RH, it would lose moisture until it reached the equilibrium value of 20%.

Under such conditions some foods may reach moisture equilibrium in the very short time of a few hours, others may require days or even weeks. When a food is in moisture equilibrium with its environment, then the a_w of the food will be quantitatively equal to the RH divided by 100.

Qualitatively, water activity is a measure of free or available water, to be distinguished from unavailable or bound water. These states of water also bear a relationship to the characteristic sigmoid shapes of water absorption isotherm curves of various foods.

Thus, according to theory, most of the water corresponding to the portion of the curve below its first inflection point is believed to be tightly bound water, often referred to as an adsorbed mono-molecular layer of water. Moisture corresponding to the region above this point and up to the curve's second inflection point is thought to exist largely as multi-molecular layers of water less tightly held to food constituent surfaces. Beyond this second inflection point moisture generally is considered to be largely free water condensed in capillaries and interstices within the food. In this latter portion of the sorption isotherm curve small changes in moisture content result in great changes in a food's a_w.

Preservation by Drying/dehydration

The technique of drying is probably the oldest method of food preservation practiced by mankind. The removal of moisture prevents the growth and reproduction of micro-organisms causing decay and minimises many of the moisture mediated deterioration reactions. It brings about substantial reduction in weight and volume minimising packing, storage and transportation costs and enable storability of the product under ambient temperatures, features especially important for developing countries. The sharp rise in energy costs has promoted a dramatic upsurge in interest in drying world-wide over the last decade.

Heat and Mass Transfer

Dehydration involves the application of heat to vaporise water and some means of removing water vapour after its separation from the fruit/vegetable tissues. Hence it is a combined/simultaneous (heat and mass) transfer operation for which energy must be supplied.

A current of air is the most common medium for transferring heat to a drying tissue and convection is mainly involved.

The two important aspects of mass transfer are:

- the transfer of water to the surface of material being dried and
- the removal of water vapour from the surface.

In order to assure products of high quality at a reasonable cost, dehydration must occur fairly rapidly. Four main factors affect the rate and total drying time:

- the properties of the products, especially particle size and geometry;
- the geometrical arrangement of the products in relation to heat transfer medium (drying air);
- the physical properties of drying medium/ environment;
- the characteristics of the drying equipment.

It is generally observed with many products that the initial rate of drying is constant and then decreases, sometimes at two different rates. The drying curve is divided into the constant rate period and the falling rate period.

Surface area

Generally the fruit and vegetables to be dehydrated are cut into small pieces or thin layers to speed heat and mass transfer. Subdivision speeds drying for two reasons:

- large surface areas provide more surface in contact with the heating medium (air) and more surface from which moisture can escape;
- smaller particles or thinner layers reduce the distance heat must travel to the centre of the food and reduce the distance through which moisture in the centre of the food must travel to reach the surface and escape.

Temperature

The greater the temperature difference between the heating medium and the food the greater will be the rate of heat transfer into the food, which provides the driving force for moisture removal. When the heating medium is air, temperature plays a second important role.

As water is driven from the food in the form of water vapour it must be carried away, or else the moisture will create a saturated atmosphere at the food's surface which will slow down the rate of subsequent water removal. The hotter the air the more moisture it will hold before becoming saturated.

Thus, high temperature air in the vicinity of the dehydrating food will take up the moisture being driven from the food to a greater extent than will cooler air. Obviously, a greater volume of air also can take up more moisture than a lesser volume of air.

Air velocity

Not only will heated air take up more moisture than cool air, but air in motion will be still more effective. Air in motion, that is, high velocity air, in addition to taking up moisture will sweep it away from the drying food's surface, preventing the moisture from creating a saturated atmosphere which would slow down subsequent moisture removal. This is why clothes dry more rapidly on a windy day.

Dryness of air

When air is the drying medium of food, the drier the air the more rapid is the rate of drying. Dry air is capable of absorbing and holding moisture. Moist air is closer to saturation and so can absorb and hold less additional moisture than if it were dry. But the dryness of the air also determines how low a moisture content the food product can be dried to.

Atmospheric pressure and vacuum

If food is placed in a heated vacuum chamber the moisture can be removed from the food at a lower temperature than without a vacuum. Alternatively, for a given temperature, with or without vacuum, the rate of water removal from the food will be greater in the vacuum. Lower drying temperatures and shorter drying times are especially important in the case of heat-sensitive foods.

Evaporation

As water evaporates from a surface it cools the surface. The cooling is largely the result of absorption by the water of the latent heat of phase change from liquid to gas.

In doing this the heat is taken from the drying air or the heating surface and from the hot food, and so the food piece or droplet is cooled.

Time and temperature

Since all important methods of food dehydration employ heat, and food constituents are sensitive to heat, compromises must be made between maximum possible drying rate and maintenance of food quality.

As is the case in the use of heat for pasteurization and sterilisation, with few exceptions drying processes which employ high temperatures for short times do less damage to food than drying processes employing lower temperatures for longer times.

Thus, vegetable pieces dried in a properly designed oven in four hours would retain greater quality than the same products sun dried over two days.

Several drying processes will achieve dehydration in a matter of minutes or even less if the food is sufficiently subdivided.

Drying Techniques

Several types of dryers and drying methods, each better suited for a particular situation, are commercially used to remove moisture from a wide variety of food products including fruit and vegetables.

While sun drying of fruit crops is still practiced for certain fruit such as prunes, figs, apricots, grapes and dates, atmospheric dehydration processes are used for apples, prunes, and several vegetables; continuous processes as tunnel, belt trough, fluidised bed and foam-mat drying are mainly used for vegetables.

Spray drying is suitable for fruit juice concentrates and vacuum dehydration processes are useful for low moisture / high sugar fruits like peaches, pears and apricots.

Factors on which the selection of a particular dryer/ drying method depends include:

- form of raw material and its properties;
- desired physical form and characteristics of dried product;
- necessary operating conditions;
- operation costs.

There are three basic types of drying process:

- sun drying and solar drying;
- atmospheric drying including batch (kiln, tower and cabinet dryers) and continuous (tunnel, belt, belt-trough, fluidised bed, explosion puff, foam-mat, spray, drum and microwave);
- sub-atmospheric dehydration (vacuum shelf/belt/drum and freeze dryers).

The scope has been expanded to include use of low temperature, low energy process like osmotic dehydration.

As far dryers are concerned, one useful division of dryer types separates them into air convection dryers, drum or roller dryers, and vacuum dryers. Using this breakdown, Table 1 indicates the applicability of the more common dryer types to liquid and solid type foods.

Table 1. Common dryer types used for liquid and solid foods

Dryer type	*Usual food type*
Air convection dryers	
kiln	pieces
cabinet, tray or pan	pieces, purées, liquids
tunnel	pieces
continuous conveyor belt	purées, liquids
belt trough	pieces
air lift	small pieces, granules
fluidized bed	small pieces, granules
spray	liquid, purées
Drum or roller dryers	
atmospheric	purées, liquids
vacuum	purées, liquids
Vacuum dryers	
vacuum shelf	pieces, purées, liquids
vacuum belt	purées, liquids
freeze dryers	pieces, liquids

Natural Drying

Surplus production and specifically grown crops may be preserved by natural drying for use until the next crop can be grown and harvested. Natural dried products can also be transported cheaply for distribution to areas where there are permanent shortages of fruit and vegetables.

The methods of producing sun and solar dried fruit and vegetables described here are simple to carry out and inexpensive. They can be easily employed by grower, farmer, cooperative, etc.

The best time to preserve fruits and vegetables is when there is a surplus of the product and when it is difficult to transport fresh materials to other markets. This is especially true for crops which are very easily damaged in transport and which stay in good condition for a very short time. Preservation extends the storage (shelf) life of perishable foods so that they can be available throughout the year despite their short harvesting season.

Sun and solar drying of fruits and vegetables is a cheap method of preservation because it uses the natural resource/ source of heat: sunlight. This method can be used on a commercial scale as well at the village level provided that the climate is hot, relatively dry and free of rainfall during and immediately after the normal harvesting period. The fresh crop should be of good quality and as ripe (mature) as it would need to be if it was going to be used fresh. Poor quality produce cannot be used for natural drying.

Dried fruit and vegetables have certain advantages over those preserved by other methods. They are lighter in weight than their corresponding fresh produce and, at the same time, they do not require refrigerated storage. However, if they are kept at high temperatures and have a high moisture content they will turn brown after relatively short periods of storage.

Different lots at various stages of maturity (ripeness) must NOT be mixed together; this would result in a poor dried product. Some varieties of fruit and vegetables are better for natural drying than other; they must be able to withstand natural drying without their texture becoming tough so that they are not difficult to reconstitute. Some varieties are unsuitable because they have irregular shape and there is a lot of wastage in trimming and cutting such varieties.

Damaged parts which have been attacked by insects, rodents, diseases, etc. and parts which have been discoloured or have a bad appearance or colour, must be removed. Before trimming and cutting, most fruit and vegetables must be washed in clean water. Onions are washed after they have been peeled.

Trimming includes the selection of the parts which are to be dried, cutting off and disposing of all unwanted material. After trimming, the greater part of the fruit and vegetables cut into even slices of about 3 to 7 mm thick or in halves/quarters, etc.

It is very important to have all slices/parts in one drying lot of the same thickness/size; the actual thickness will depend on the kind of material. Uneven slices or different sizes dry at different rates and this result in a poor quality end product. Onions and root crops are sliced with a hand slicer or vegetable cutter; bananas, tomatoes and other vegetables or fruit are sliced with stainless-steel knives.

As a general rule plums, grapes, figs, dates are dried as whole fruits without cutting/slicing.

Some fruit and vegetables, in particular bananas, apples and potatoes, go brown very quickly when left in the air after peeling or slicing; this discoloration is due to an active enzyme called phenoloxidase. To prevent the slices from going brown they must be kept under water until drying can be started. Salt or sulphites in solution give better protection. However, whichever method is used, further processing should follow as soon as possible after cutting or slicing.

Blanching

Blanching—exposing fruit and vegetable to hot or boiling water—as a pre-treatment before drying has the following advantages:

- it helps clean the material and reduce the amount of micro-organisms present on the surface;
- it preserves the natural colour in the dried products; for example, the carotenoid (orange and yellow) pigments dissolve in small intracellular oil drops during blanching and in this way they are protected from oxidative breakdown during drying;
- it shortens the soaking and/or cooking time during reconstitution.

During hot water blanching, some soluble constituents are leached out: water-soluble flavours, vitamins (vitamin C) and sugars. With potatoes this may be an advantage as leaching out of sugars makes the potatoes less prone to turning brown.

Blanching is a delicate processing step; time, temperature and the other conditions must be carefully monitored.

A suitable water-blanching method in traditional processing is as follows:

- the sliced material is placed on a square piece of clean cloth; the corners of the cloth are tied together;
- a stick is put through the tied corners of the cloth;
- the cloth is dipped into a pan containing boiling water and the stick rests across the top of the pan thus providing support for the cloth bag.

The average blanching time is 6 minutes. The start of blanching has to be timed from the moment the water starts to boil again after the cloth bag has been dipped into the pan. While the material is being blanched the cloth bag should be raised and lowered in the water so that the material is heated evenly.

When the blanching time is completed the cloth bag and its content should be dipped into cold water to prevent over-blanching. If products are over-blanched (boiled for too long) they will stick together on the drying trays and they are likely to have a poor flavour.

Green beans, carrots, okra, turnip and cabbage should always be blanched. The producer can choose whether or not potatoes need blanching. Blanching is not needed for onions, leeks, tomatoes and sweet peppers. Tomatoes are dipped into hot water for one minute when they need to be peeled but this is not blanching.

Preservatives are used to improve the colour and keeping qualities of the final product for some fruits and vegetables. Preservatives include items such as sulphur dioxide, ascorbic acid, citric acid, salt and sugar and can either be simple or compound solutions.

Treatment with preservatives takes place after blanching or, when blanching is not needed, after slicing. In traditional, simple processing the method recommended is:

- put enough preservative solution to cover the cloth bag into a container/pan;
- dip the bag containing the product into the preservative solution for the amount of time specified;
- remove the bag and put it on a clean tray while the liquid drains out. The liquid which drains out must not go back into the preservative solution because it would weaken the solution.

Care must be taken after each dip to refill the container to the original level with fresh preservative solution of correct strength. After five lots of material have been dipped, the remaining solution is thrown away; i.e. a fresh lot of preservative solution is needed for every 5 lots of material. The composition and strength of the preservative solution vary for different fruit and vegetables.

The strength of sulphur dioxide is expressed as "parts per million" (ppm). 1.5 grams of sodium metabisulphite in one litre of water gives 1000 ppm of sulphur dioxide. Details for solutions of different strengths are given in table 2.

Sodium bicarbonate is added to the blanching water when okra, green peas and some other green vegetables are blanched. The chemical raises the pH of the blanching water and prevents the fresh green colour of chlorophyll being changed into pheophytin which is unattractive brownish-green.

The preservative solutions in the fruit and vegetable pre-treatment can only be used in enamelled, plastic or stainless-steel containers; never use ordinary metal because solutions will corrode this type of container.

As a general rule, preservatives are not used for treating onions, garlic, leeks, chilies and herbs.

Table 2. Dilutions of sodium metabisulphite with water to obtain "PM" of sulphur dioxide (SO_2)

PPM SO_2	*Sodium Metabisulphite*	
	Grams per litre of water	*Grams per 20 litre tin of water*
1000	1.5	30.0
2000	3.0	60.0
3000	4.5	90.0
4000	6.0	120.0
5000	7.5	150.0
6000	9.0	180.0
7000	10.5	210.0

One level teaspoon of sodium metabisulphite = c. 5 g.

Osmotic Dehydration

In osmotic dehydration the prepared fresh material is soaked in a heavy (thick liquid sugar solution) and/or a strong salt solution and then the material is sun or solar dried. During osmotic treatment the material loses some of its moisture. The syrup or salt solution has a protective effect on colour, flavour and texture. This protective effect remains throughout the drying process and makes it possible to produce dried products of high quality. This process makes little use of sulphur dioxide.

Sun Drying

The main problems for sun drying are dust, rain and cloudy weather. Therefore, drying areas should be dust-free and whenever there is a threat of a dust storm or rain, the drying trays should be stacked together and placed under cover.

In order to produce dust-free and hygienically clean products, fruit and vegetable material should be dried well above ground level so that they are not contaminated by dust, insects, livestock or people. All materials should be dried on trays designed for the purpose; the most common drying trays have wooden frames with a fitted base of nylon mosquito netting. Mesh made of woven grass can also be used. Metal netting must NOT be used because it discolours the product.

The trays should be placed on a framework at table height from the ground. This allows the air to circulate freely around the drying material and it also keeps the food product well away from dirt. Ideally the area should be exposed to wind and this speed up drying, but this can only be done if the wind is free of dust.

With 80 cm x 50 cm trays, the approximate load for a tray is 3 kg; the material should be spread in even layers. During the first part of the drying period, the material should be stirred and turned over at least once an hour.

This will help the material dry faster and more evenly, prevent it sticking together and improve the quality of the finished product. Products for sun drying should be prepared early in the day; this will ensure that the material enjoys the full effect of the sun during the early stages of drying.

At night the trays should be stacked in a ventilated room or covered with canvas. Plastic sheets should NEVER be used for covering individual trays during sun drying.

Dry or nearly dry products can be blown out of the tray by the wind. However, this can be protected by covering the loaded tray with an empty one; this also gives protection against insects and birds.

Shade Drying

Shade drying is carried out for products which can lose their colour and/or turn brown if put in direct sunlight. Products which have naturally vivid colours like herbs, green and red sweet peppers, chilies, green beans and okra give a more attractive end-product when they are dried in the shade.

The principles for the shade drying are the same as for sun drying. The material to be dried requires full air circulation. Therefore, shade drying is carried out under a roof or thatch which has open sides; it cannot be done either inside conventional buildings with side walls or in compounds sheltered from wind. Under dry conditions when there is a good circulation of air, shade drying takes little more time than is normally required for drying in full sunlight.

Solar Dryers and their Applications

In the selection of appropriate solar dryers for commercial scale operation, it is imperative that economics be kept in view at all time. A total Energy System concept should be employed and due consideration be given to parasitic energy consumption.

The following features have been identified:

- large scale dryers are more promising than small scale ones. However, small scale dryers should not be neglected.
- the dryer should be designed to maximise the utilisation factor of the capital investment, i.e. multi-products (fruit, vegetables and other raw material) and multi-use (e.g. drying and heating water for domestic use).
- in general, an auxiliary heat source should be provided to assure reliability, to handle peak loads and also to provide continuous drying during periods of no sunshine.
- forced convection indirect dryers are preferred because they offer better control, more uniform drying and because of their high heat collection efficiency result in smaller collector area. However, parasitic power should be kept to a minimum.

Two dryer systems have been identified:

- a cabinet type dryer with natural convection for internal air circulation for the processing of dried fruit such as mango, banana, pineapple, apricot, pear, apple, etc. and also for potato chips and other vegetables;
- a greenhouse type dryer with forced air circulation.

Some of the barriers to the commercial development of solar dryers have been attributed to:

- initial cost - poor farmers cannot afford them
- durability - constant breakdown due to using low cost building materials
- misuse - through lack of training and technical skills
- dependability and reliability - during the wet season when drying is critical there is not enough solar energy available
- the wider use of Solar Drying Systems has been limited by other factors which are not necessarily of a technical or technological nature. Among the most important are the lack of national policies directed to promoting the drying of produce at the production site, in order to reduce losses, improve quality and increase farmers' earnings.

Construction of Solar Dryers

In the case of simple natural convection dryers it may be more appropriate to build and operate a number of small units. Multiplicity allows diversity, since more than one crop can be dried at a time. A further advantage is that if one dryer is out of operation due to damage, drying can still continue at reduced capacity using the other dryers.

On the other hand, more sophisticated dryers, such as forced convection solar dryers, benefit from economies of scale due to the investment tied up in the fan and the source of heat.

Generally speaking, one large dryer will be more cost-effective than two smaller units. However, it should be taken into consideration that an oversized unit will be operating at less than full capacity, reducing any cost advantage. The drying area required will depend on local conditions, commodity and number of trays on each rack or trolley.

Construction methods and materials

Construction methods and available materials may vary considerably from location to location. Some general guidelines regarding factors which must be considered can, however, be given:

— *dimensions of standard materials*. Where possible, design should take account of the sizes of material locally available. For example, it would be poor design to specify the width of a corrugated iron collector as 1.1 m if the standard width of a corrugated iron sheet is 1 m.

  Before finalising a design the commercial availability of materials must be ascertained.

— *use of rural materials*. The cost of building of solar dryer can be minimised if the producer is able to use wood cut straight from the forest rather than prepared timber.

Careful design in the development stage of a dryer can often facilitate the use of cheaper materials. Difficulties caused by these materials are in joining pieces of the structure, in sealing the structure against air leaks, and in attaching the plastic sheet to the frame. There is obvious scope for designs which use prepared timber for strategic points and unprepared at others.

Where the use of wood is necessary, remember to take environmental factors into consideration. For example, determine the effect of flash flooding or termites might be and take the appropriate preventive action.

— *use of plastic sheets*. For many solar dryers, the clear plastic sheet used is the major capital cost to the farmer; therefore, the type of plastic chosen is important.

A choice must be made between a relatively cheap plastic such as ordinary polyethylene which will last, at best, for one season due to photo-degradation and wear and tear; and a more expensive, better quality plastic less prone to photo-degradation; or even glass or a rigid plastic.

Attaching plastic sheet to the framework structure, so as to minimise the likelihood of the plastic being torn is, perhaps, the most difficult part of building a dryer. Listed below are some general points which should be followed to prolong the useful lifetime of plastic sheet on a solar dryer:

- when attaching plastic sheet to the framework, care should be taken to stretch the plastic at the points of attachment, but the plastic should not be so loose that it will flap about in the wind;
- rather than merely stapling or nailing the plastic directly to the framework, it is preferable to sandwich the plastic between the framework and a batten. This may not be practical when unprepared wood or other materials are being used.
- no sharp edges should come in contact with the plastic sheet since these will initiate tears;
- fold over the plastic at the point of attachment to the frame, so that there are two or more layers of plastic. This will help prevent tears;
- when fixing the sheet over the framework, sags and hollows in which water can collect should be avoided wherever possible;
- the dryer should be handled as carefully and as seldom as possible during operation and when not in use.

5.2.5.2 Technical criteria

The following design factors must be established:

- the throughput of the dryer over the productive season; the size of batch to be dried;
- the drying period(s) under stated conditions;
- the initial and desired final moisture content of the commodity (if known);
- the drying characteristics of the commodity, such as maximum drying temperature, effect of sunlight upon the product quality, etc.;

- climatic conditions during the drying season, i.e. sunlight intensity and duration; air temperature and humidity; wind speed (such data may be available from local meteorological stations);
- availability and reliability of electrical power;
- the availability, quality, durability and price of potential construction materials such as:
  - glazing materials: glass, plastic sheet or film;
  - wood (prepared or unprepared);
  - nails, screw, bolts, etc.;
  - metal sheet, flat or corrugated angle iron;
  - bricks (burnt or mud), concrete blocks, stones, cement, sand, etc.
  - roofing thatch;
  - metal mesh, wire netting, etc.
  - mosquito netting, muslin, etc.
  - bamboo or fibre weave;
  - black paint, other blackening materials;
  - insulation material; sawdust, etc.;
- the type of labour available to build and operate the dryer;
- the availability of clean water at the site for preparation of the commodity prior to drying.

In any one situation there may well be other technical factors that need to be considered.

Socio-economic criteria

From the initial considerations, estimates of the capital costs of the dryer, the price of the commodity to be dried, and the likely selling price of the dried product will have been made. Other question that need to be considered are the following:

- who will own the dryer?
- is the dryer to be constructed by the end-user (with or without advice from extension agencies), local contractors, or other organisations?
- who will operate and maintain it?
- how can the drying operation be incorporated into current practices?

- are sources of finance from local authorities or extension agencies available, etc.?

Obviously there are many other socio-economic factors, particularly those of a local nature, which must be taken into account. It cannot be stressed too highly that if such factors are not taken into account and evaluated, then is every chance that an inappropriate dryer design may result. Equal emphasis must be placed on both technical and socio-economic factors.

Sun/solar Drying Tray

The drying tray described requires seasoned timber 22.5 mm thick x 50 mm wide.

A sun drying tray requires 6 meters of seasoned timber 22.5 mm thick x 50 mm wide.

Cut the timber into lengths of 900 mm long for the sides of the tray and 600 mm long for the ends - 4 pieces of each length will be needed. The ends of each piece are cut as shown in the drawing - this is to make flush fitting joints. Join the corners using small brass screws 20 mm long. To make extra strong joints use good quality wood glue as well as the screws.

The nylon mosquito netting or grass woven mesh can be fitted between the frames as shown in the bottom drawing. Cut the mesh a little larger than the size of the frame. Using drawing pins, pin the mesh to the OUTSIDE edges of one of the frames - the mesh should be pulled tight as the pins are put in around the edges.

Lay the other frame on top and drill holes about 3 mm in diameter at the points marked X in the top drawing. Use nails that are a tight fit in the holes and tap gently into place leaving a portion standing above the frame.

Cut off the standing part to leave a piece about 12 mm long which is then bent over and tapped firmly down onto the frame. When the frame has been put together tightly, the drawing pins can be removed.

Preservation by Concentration

Foods are concentrated for many of the same reasons that they are dehydrated; concentration can be a form of preservation but this is true only for some foods. Concentration reduces weight and volume and results in immediate economic advantages.

Nearly all liquid foods which are dehydrated are concentrated before they are dried. This is because in the early stages of water removal, moisture can be more economically removed in highly efficient evaporators than in dehydration equipment. Further, increased viscosity from concentration often is needed to prevent liquids from running off drying surfaces or to facilitate foaming or puffing.

Foods are also concentrated because the concentrated forms have become desirable components of diet in their own right. Thus, fruit juices plus sugar with concentration becomes jelly. The more common concentrated fruit and vegetable products include items as fruit and vegetable juices and nectars, jams and jellies, tomato paste, many types of fruit purées used by bakers, candy makers and other food manufacturers.

Aspects of Preservation by Concentration

The level of water in virtually all concentrated foods is in itself more than enough to permit microbial growth. Yet while many concentrated foods such as non-acid fruit and vegetable purées may quickly undergo microbial spoilage unless additionally processed, such items as sugar syrups, jellies and jams are relatively "immune" to spoilage; the difference of course is in what is dissolved in the remaining water and what osmotic concentration is reached.

Sugar and salt in concentrated solutions have high osmotic pressure. When these are sufficient to draw water from microbial cells or prevent normal diffusion of water into these cells, a preservative condition exists.

The critical concentration of sugar in water to prevent microbial growth will vary depending upon the type of micro-organisms and the presence of other food constituents, but usually 70% sucrose in solution will stop growth of all micro-organisms in foods. Less than this concentration may be effective but for short periods of time unless the foods contain acid or they are refrigerated.

Salt becomes a preservative when its concentration is increased and levels of about 18% to 25% in solution generally will prevent all growth of micro-organisms in foods.

Except in the case of certain briny condiments, however, this level is rarely tolerated in foods.

Removal of water by concentration also increases the level of food acids in solution (particularly significant in concentrated fruit juices).

Reduced Weight and Volume by Concentration

While the preservation effects of food concentration are important, the main reason of most food concentration is to reduce food weight and bulk. Tomato pulp which is ground tomato minus the skins and seeds, has a solid content of only 6 % and so a 3.785 litre can would contain only 231 g of tomato solids.

Table 3. Specific gravity and solids. Tomato pulp and commercial tomato concentrates

Tomato Solids, %	*Specific gravity at 68°F (20°C)*	*Dry Tomato Solids per Gal. at 68°F, lb*	*per Litre at 20°C, g*
6.0 Tomato pulp	1.025	0.51	61
10.8	1.045	0.94	113
12.0 Tomato purée	1.050	1.05	126
14.2	1.060	1.25	151
16.5	1.070	1.47	177
25.0	1.107	2.31	277
26.0	1.112	2.41	289
28.0 Tomato paste	1.120	2.61	314
30.0	1.129	2.82	339
32.0	1.138	3.03	364

Concentrated to 32% solids, the same can would contain 1.38 kg of tomato solids or six times the value of product. For a manufacturer needing tomato solids such a producer of soups, canned spaghetti or frozen pizza the saving from concentration are enormous in cans, transportation costs, warehousing costs and handling costs throughout his operation.

Methods of Concentration

a. *Solar concentration.* As in food dehydration, one of the simplest methods of evaporating water is with solar energy. A typical example of this method is production at farm level in developing countries of fruit pastes/leathers (such as apricot or plum pastes).

b. *Open Kettles.* Some foods can be satisfactorily concentrated in open kettles that are heated by steam. This is the case for jellies and jams, tomato juices and purées and for certain types of soups. High temperatures and long concentration times should be avoided in order to reduce or eliminate damage. It is also necessary to avoid thickening and burn-on of product to the kettle wall as these gradually lower the efficiency of heat transfer and slow the concentration process.

c. However, when the process is under control, this type of evaporation is still highly recommended for small scale operations in developing countries. It is a quite widely used system, mainly for jellies, jams and marmalades.

d. Flash evaporators.

e. Thin-film Evaporators.

f. *Vacuum evaporators.* - It is common to construct several vacuumised vessels in series so that the product moves from one vacuum chamber to the next and thereby becomes progressively more concentrated in stages.

g. With such an arrangement the successive stages are maintained at progressively higher degrees of vacuum, and the hot water vapour produced by the first stage is used to heat the second stage, the vapour from the second stage heats the third stage and so on. In this way maximum use of heat energy is made. .

h. *Freeze Concentration.* This process has been known for many years and has been applied commercially to orange juice. However, high processing costs due largely to losses of juice occlude [unclear] to the ice crystals, have limited the number of installations to date.

Changes from Concentration

Obviously concentration that exposes food to 100° C or higher temperatures for prolonged periods can cause major changes in organoleptic and nutritional properties. Cooked foods and darkening of colour are two of the more common heat induced results which must be kept under control during a well designed process with an efficient evaporator which is still "safe".

Microbial destruction is another type of change that may occur during concentration and will be largely dependent upon temperature. Concentration at a temperature of 100 °C or slightly above will kill many micro-organisms but cannot be depended on to destroy bacterial spores. When the food contains acid, such as fruit juices, the kill will be greater but again sterility is unlikely.

On the other hand, when concentration is done under vacuum many bacterial types not only survive the low temperatures but multiply in the concentrating equipment. It is therefore necessary to stop frequently and sanitise low temperature evaporators and where sterile concentrated foods are required, to resort to an additional preservation treatment.

References

Brett, A et al. 1996. *Producing Solar Dried Fnlits and Vegetables jor Micro- and Small-scale Rural Enterprise Development: A Series of Practical Guides.* Chatham, UK: NRI

DeLong, D. 1979. *How to Dry Foods.* Los Angeles: HP Books.

Hobson, P. 1994. *Making and Using Dried Foods,* Pownal, VT: Storey Communications, Inc.

Miller, M. et al. 1981. *Drying Foods at Home.* University of California, Division of Agricultural Science, Leaflet 2785.

8

Processing for Home Consumption

The majority of fruits and vegetables grow in short seasons and one main reason for processing is therefore to secure a supply of these foods when they are not available in a fresh state. This is both because some (tomatoes, onions, spices etc.) are used in daily cooking in different societies and because regular consumption of fruits and vegetables is necessary to maintain health.

In different societies throughout the world, the consumption of fruits and vegetables has different levels of significance in the diet. In some areas, such as the cold, infertile conditions of Central Asian steppes, mountainous regions of Latin America and arid deserts of Africa and Asia, fruit and vegetable cultivation is difficult and has a short annual growing season. These products do not therefore play an important role in local diets and there is greater reliance on animal products. However, the requirement to eat fruit and vegetable products to maintain health means that in almost all of these societies, small amounts of leafy vegetables and collected fruits are preserved for winter or dry seasons and as a result of their shortage, they have a high value.

By contrast, in almost every country of the humid tropics and sub-tropical regions, the soils and climate are suitable for fruit and vegetable cultivation and there is an historic reliance on these products as an important part of the normal daily diet. Indeed in some areas this may be reinforced by religious beliefs, such as Hinduism, which requires followers to be vegetarians. In many communities throughout the world, vegetables are eaten daily as an accompaniment to a main cereal or root crop meal, with or without meat and fish. Fruits are eaten fresh when in season as appetisers, snacks, juices and desserts.

There are a large number of traditional processed fruit and vegetable products that are traditionally made in the home. Examples from Asia include fruit 'leathers' (dried sheets of fruit paste that have an appearance of thin leather), fruit pastes, pickles and chutneys. In Africa and Latin America, a wide range of dried chips, dried and powdered leaves, fruit beers and other fermented fruit and vegetable products are traditionally produced. Each of these products is stored for future use and is often not intended for sale, although no clear distinction is possible. Some products that are processed for home consumption may also be sold to neighbours or at village markets when additional income is needed.

In general terms because of low incomes, home processing requires simple, cheap equipment and will usually be done without specialised machines, using existing kitchen utensils and simple pieces that can be made from local materials at almost no cost. Processing methods are also inexpensive but in some cases are complex and sophisticated, involving a sequence of process stages that require considerable experience and expertise to perform correctly.

Some of the main reasons for home processing are to maintain a secure supply of food for seasons of shortage (to provide food security) and to maintain health through an adequate supply of nutrients in the diet.

Food Security, Nutrition and Health

Fruits and vegetables provide an abundant, cheap source of fibre and several vitamins and minerals. In general, they have the highest nutritional value when eaten fresh, although an exception may be fermented foods, in which the process of fermentation can increase the content of B-vitamins.

However, fresh fruits and vegetables have a relatively short life before they begin to decay and if they are to be stored for an extended period, it is necessary to process them by one or more of the various methods that are available. The extent to which preservation is required depends in part on the type of plant material that is being considered. For example, fast-growing parts such as shoots have a short life before rotting, whereas bulbs and tubers can be stored without preservation for much longer at room temperature (25°C).

The extent to which nutrients are lost during processing varies according to both the type of fruit or vegetable and the process that is used.

The purposes of processing fruits and vegetables in the home may therefore be summarised as follows:

- preservation of seasonal gluts for survival in times of shortage
- provision of cooking ingredients throughout the year
- increase in convenience by having part-prepared foods available when required.
- availability of different foods throughout the year to provide variety and greater nutritional balance in the diet.

Table 1. Dietary fibre, selected vitamins and minerals in fruits and vegetables compared with some other foods.

Food	*Fibre (g/100g)*	*Carotene (mg/100g)*	*Thiamine (mg/100g)*	*Ascorbic acid (Vitamin C) (mg/100g) (mg/100g)*	*Calcium (mg/100g)*	*Iron*
Soya flour	11.9	-	0.75	0	210	6.9
Wholemeal bread	8.5	0	0.26	0	23	2.5
Boiled rice	2.4	0	0.01	0	1	0.2
Doughnut	-	0	-	-	70	1.9
Whole goats' milk	0	22	0.04	1.5	130	0.04
Yoghurt	0	5	0.05	0.4	180	0.09
Eggs (raw)	0	tr	0.09	0	52	2.0
Beef (raw)	0	tr	0.05	0	7	1.9
Chicken (raw)	0	tr	0.10	0	10	0.7
Green beans (raw)	2.9	400	0.05	20	27	0.8
Green cabbage (boiled)	2.2	500	0.03	25	30	0.5
Carrot (raw)	2.9	12,000	0.06	6	48	0.6
Okra (raw)	1.3	90	0.10	25	70	1.0
Plantain (raw)	5.8	60	0.05	20	7	0.5
Tomato (raw)	1.5	600	0.06	20	13	0.4
Apricot (raw)	2.1	1,500	0.04	7	17	0.4
Banana (raw)	3.4	200	0.04	10	7	0.4
Mango (raw)	1.5	1,200	0.03	30	10	0.5
Orange (raw)	2.0	38	0.08	38	31	0.3
Pineapple (raw)	1.2	60	0.08	25	12	0.4

tr: = trace amount

Table 2. Rate of spoilage of different parts of fruit and vegetable plants

Part of plant	*Time before spoilage starts (days)*
Fast growing shoots	1-2
Leaves	2-3
Stems	5-50
Fruits	10-30
Roots and tubers	15-50
Bulbs	30-100

When the objective of a programme is to improve health and nutritional status by increasing the effectiveness of food processing, it is usually necessary to combine education and training with practical improvements to facilities and equipment. In order for people to learn from each others' experience and to make such interventions cost effective, it is usually necessary for participants to organise themselves into groups. This in itself can be a long and difficult process in some communities, but where groups such as mothers' clubs or farmers' societies already exist, they can be approached to act as a focus for a new programme. In some instances, successful food processing developments for improved nutrition have also been linked to health centres.

Any improvements to facilities and equipment need to be low cost and easily maintained if they are to be used in a sustained way. Whereas the capital cost of a new building or item of processing equipment can be donated, the facility will cease to operate if both funds and responsibility for maintenance are not established at the outset. Many attempts have been made to share the costs of processing by establishing various forms of producer associations or village co-operatives, with varying degrees of success.

One important factor is the degree of trust and willingness to work together that exists in a community. If there is no history of a communal approach to problem solving or implementing a new development, it can take many years to introduce these ideas. However, there are many examples of successful village associations that have been operating for several years and have supported on-going education and training programmes in health and nutrition, together with the establishment of community systems and facilities for crop processing.

Improvements to Home Processing and Storage

In most cases, simple methods of processing and storage that are based on traditional knowledge have the greatest chance of being successfully introduced. Improvements to existing practices are more likely to succeed than completely new technologies, not only because they are more familiar to people, but also because traditional methods are suited to local environmental conditions, whereas new technologies may not be. A good example of such an approach is improved drying and storage.

Drying

Drying is one of the most accessible and hence the most widespread processing technology. When combined with improved food stores, it can lead to significant improvements in food security in most regions. Important examples of dried fruits and vegetables are okra, cabbage, spinach, mango slices or leathers, garlic and other flavourings, all of which are stored for use in daily cooking.

Sun drying is traditionally practised because there is negligible cost in processing and the work of spreading and turning the crop as it dries can be shared among family members, including young children. Some of the reasons why the introduction of solar dryers into village communities has not been widely successful concern these factors of cost and workload. Even when constructed from locally available materials, the simplest dryer still has a higher cost than sun drying. Additionally, the work involved in loading and carrying trays of fruits and vegetables and removing them regularly to turn and mix the products, is not suitable for children and therefore places an additional workload on the parents, usually the mothers. Furthermore, the possibility of improved quality products and shorter drying times using solar dryers may not be important factors to families who may be satisfied with the existing quality of dried foods and the time and effort required to dry them.

Under such circumstances, a focus on improvements to sun drying is likely to yield greater results. Examples include raising crops off the ground on woven matting or wooden drying frames to reduce contamination by dust, crawling insects or rodents; covering the crop with mosquito netting to reduce damage and contamination from birds and flying insects; cutting the fruits and vegetables into smaller pieces so that they dry faster and blanching vegetables to retain a better colour. Although it is beyond the scope of this book, it may be possible in some programmes to link drying improvements to agricultural developments and the introduction of crop varieties that are better suited to drying than traditional types.

Concentration by Boiling

Relatively few fruits and vegetables are preserved in the home by concentration, mainly because of the additional fuel-wood consumption, but important foods in some societies are tomato paste, extracts from wild plants such as baobab fruit, that are used as flavourings, chutneys and syrups made from fruits or saps. The main requirement to improve processing of these products is to control the rate of heating to prevent localised burning of the product, particularly when it has become thickened towards the end of boiling.

Fermentation

Production of fruit wines and beers is possibly the second most common form of home processing after drying, in countries where alcohol consumption is not prohibited by religious beliefs. A wide range of fruits is used, particularly pineapple, melon and pawpaw. In the example below from East Africa, banana is used. It is difficult to produce a clear juice from bananas but the following description by the processor indicates the complexity and degree of skill required to produce a satisfactory product.

Bananas are harvested when they are not quite ripe and not over-ripe. They are stored in a pit covered with banana leaves for up to 4 days to ripen. If necessary a fire is lit in a small hole leading into the pit to warm the bananas and hasten ripening. The bananas are peeled and placed into a large wooden trough. They are then crushed to a pulp by walking on the fruit. Two grasses, one of which is young elephant grass, are cut to approximately 15 cm long and carefully mixed in the correct proportions. Thin layers are periodically added to the surface as pulping proceeds and the grass becomes mixed into the pulp. This is to finely divide the pulp into small particles and thus release the juice. Only these two grasses are used as they do not add flavour to the juice in the way that other species do.

After approximately 30-45 minutes pulping, the mixed grass and pulp is piled on a coarse filter frame, made from wooden poles lashed together over the trough, and juice is pressed out by standing on the pulp. The juice is then collected using a calabash into which a hole is cut in one side and a hollow handle is fitted to the top. Juice is poured through the hollow handle into a second calabash, which is shaped as a runnel and lined with grass to act as a fine filter. The juice, which is cloudy but without particles, is collected in a clean container. It is tasted at this stage and if necessary, adjusted to the correct sugar content by adding water.

The standardised juice is then returned to a clean wooden trough and sorghum is added in the ratio of 2 kg per 20 litres of juice. The trough is covered with banana leaves and also with the residue from pulping to act as insulation and maintain the temperature of fermentation at ambient (22-26°C) for two days. After fermentation, the liquor is strained through the calabash/grass filter into clean containers or bottles and is then ready for drinking. There are no further clarification or pasteurisation stages and the beer is preserved by the alcohol content (3-5%). As a result it has a shelf life of up to four days without refrigeration. The process could be improved and a longer shelf life obtained, if the product was clarified by filtration or sedimentation and pasteurised in the bottles.

Home preservation of fruits and vegetables by fermentation to produce pickles is especially common in South and Central Asia, but less so in other regions of the world. Examples include pickles from lime, mango, mixed fruit and a range of vegetables, including different types of cabbage. Each local district may have a slightly different set of ingredients or variations in the process and in India for example, there are estimated to be over 700 different types of fermented pickle.

In all cases the prepared fruit or vegetable is mixed with spices and held in a closed container such as glazed pottery, while a natural sequence of bacteria ferment sugars in the raw materials to produce lactic acid. Provided that the raw materials are not heavily infected with moulds and that insects, which can carry yeasts or moulds on their bodies, are kept out of the container, the build up of lactic acid is sufficient to prevent spoilage and the product can be kept for several months while it is being used.

In the following example from Central Asia, pickled cabbage is produced: a 2 cm layer of salt is placed into the bottom of a wide-necked pottery drum or jar. A 5 cm layer of shredded cabbage is placed onto the salt and the jar is then filled using alternating layers of salt and cabbage. It is then sealed with a wooden or cloth bung and allowed to ferment for several weeks. As the water is drawn out of the cabbage, it forms a pickling brine. If the product is not to be used straight away, the cabbage is weighted down with a clean stone to keep it beneath the brine and to prevent it drying out and spoiling. Little improvement can be made to these processes without substantially increasing capital and/or operating costs. They are well suited to local tastes and environmental conditions.

Pickling

There is also a wide range of unfermented pickles produced in Asia. Two examples below describe the production of lime pickle and mixed pickle from South Asia.

Well ripened, but not over-ripe limes are washed, cut into four (but the pieces are not separated), and dipped in a concentrated solution of salt. They are then dried in the sun for several days. Sometimes salt crystals are sprinkled onto the lime while it is drying. Drying is continued until the skin develops a brown colour and the pieces can be broken using the fingers. The product is then packed along with the expelled juice in pottery vessels. The salt removes water from the pieces due to osmosis and some sugar is also removed. As a result, salt-tolerant micro-organisms begin to grow while the product is being dried. These bacteria produce some acid, but mostly generate the characteristic flavour of the product, which has a sour and salty taste and can be kept for several months.

The main raw materials for mixed pickle are green chillies, onion, papaya, and spices. The papaya and onion are peeled and cut into slices. The spices are also washed and then ground. All ingredients are placed in a clean pot and mixed by hand while adding vinegar. The product can be kept in the sealed pot for several months, during which time there is a degree of fermentation which improves the flavour, but it does not contribute to the preservation. Other formulations of mixed pickles use raw materials such as dates, pineapple or mango.

Storage

In many cases, the most serious problems of food security are connected with inadequate food storage practices. For example, dried foods are susceptible to spoilage by mould or to attack by insects and rodents, particularly if they are inadequately dried. Losses during storage can be considerable and when these are cereal losses, they can be life threatening. Traditional storage practices for fruits and vegetables vary widely in different parts of the world, but all use locally available materials, usually at little or no cost. It is not possible in a book of this size to detail these differences, but the general principles of storage are described below, followed by some examples of good practice that may assist in improving storage practices and facilities in other places.

As described above, the main types of processed fruits and vegetables that are stored in the home are dried foods, followed by fermented and pickled foods and concentrated syrups or pastes. The latter groups are each moist foods but they are preserved by raised concentrations of lactic acid, salt or sugar. All that is required for adequate storage is therefore an impervious pot to contain the contents and an airtight and insect or rodent proof seal to prevent contamination or theft. Glazed pottery or glass vessels, with cork or wooden stoppers, sealed with resin or wax are entirely adequate for the expected shelf life of these products. They should be stored off the floor in a cool dark place, away from sunlight and dampness.

Dried fruits and vegetables need protection from insects, animals and from uptake of moisture. Those foods that are required to remain unbroken also need protection from crushing and those that contain high levels of carotene, such as green leafy vegetables or carrots, also require protection against sunlight. Because of their bulkiness, pottery vessels may be too expensive to store an adequate supply and many families resort to hanging dried fruits and vegetables from rafters or over cooking areas, where they are sure to remain dry. This deters most animals but does not provide adequate protection against insects. Improved storage can be achieved by constructing a simple mesh basket from wooden sticks and mosquito netting, which can be hung from the rafters in a cool, dry place, away from sunlight.

Dried foods should not be stored on the floor. If storage on tables or raised platforms in the house is the only option, the supporting legs should rest in small containers of oil or kerosene to deter crawling insects. Storage containers should be protected by insect proof mesh. Ideally, re-used cans, having press-on lids or similar insect- and rodent-proof containers are used, once they have been thoroughly cleaned and dried.

Home Processing to Earn Extra Family Income

There are very large numbers of people throughout the developing world who process foods in their homes and sell them to neighbours or in local markets, in order to generate a little extra money to supplement family incomes. People in this informal sector are characterised by having little capital to invest in equipment and they are forced to buy inputs such as raw materials, ingredients and packaging in small quantities from retailers, because of a lack of working capital or an uncontrolled cashflow. This increases the unit costs of their products and may make them un-competitive on price with those from the formal sector, or even in some cases with subsidised imports. They have poor access to credit because they have little or no collateral and credit providers cannot justify the high administrative costs of servicing a large number of small loans.

It is also frequently the case that informal processors have little experience of quality control and so their products do not compete well on quality, safety or appearance with those from the formal sector. Because they only process fruits or vegetables until they have earned sufficient money for their immediate needs and they then cease production until the next financial need, informal enterprises are unable to build up trusting relationships with suppliers or retailers and as a result, future orders may be more difficult to obtain.

Thus a complex mix of factors need to be addressed when a programme is designed to assist informal, home-based processors. Provision of accessible credit is one component, but by itself is insufficient to generate success. Technical training in production routines, hygiene and quality assurance is also required and most importantly, an improved ability to identify suitable markets and to develop skills in selling the products is needed.

The division of labour within the group also has social benefits not only in providing a more flexible working environment to take account of household responsibilities but also in developing, encouraging and rewarding an individual's competence and skills in their area of work. This can have dramatic effects on the confidence and determination of individuals to succeed. This approach also removes the need for each member of the group to have expertise in all areas of operating a processing enterprise (for example checking quality, negotiations with suppliers or customers, keeping records, producing financial data, marketing and selling etc.), and allows each member to develop their own specific contribution to the success of the group, based on their individual skills and abilities. There can be various structures for such groups, including producer co-operatives, in which each member has an equal share, groups based on religious beliefs, extended families or those formed around a different existing activity, such as a youth club, a mothers' union or a village development society.

A further development of group activity in the informal sector is the division of responsibility for different parts of the food supply chain. A number of groups can reach agreement to collaborate and share the profits of the joint venture. In some successful approaches, a farmers' group is responsible for supplying a local group of processors with an agreed amount of raw materials of specified quality. This removes the need for farmers to develop skills in marketing their crops and they can concentrate on their farming abilities. The processors may buy from a number of farmers in the local area, thus reducing transport costs for the raw materials. The processors concentrate on producing high quality products at the lowest cost and sell them to a marketing group. The marketing group may buy from a number of processors and concentrate only on finding the highest paying and most reliable markets for the products.

A development of this system may involve part-processing of fruits or vegetables on the farm to reduce transport costs and to add value to the farmers' produce. For example, fruits can be peeled and part-dried or crystallised on the farm and then transported to a local processor in returnable containers. There are some environmental advantages in doing this because not only are fuel costs for transportation reduced, but also localised pollution at the central processing unit is removed and fruit and vegetable wastes are available on the farm for green manure. In this system, the processor standardises the quality of each batch by blending or sorting and then packages each farmers' products under a single brand name. This gives both benefits of lower costs through bulk purchases of packaging and the ability to supply larger amounts to meet demands of retailers who would otherwise not consider purchasing.

References

Colin, D. 1992. Recent trends in fruit and vegetable processing. In *Food Science and Technology Today*, 7, (2)), pp. 111 - 116.

Desrosier, N.W. and Desrosier, J.N. 1977. *Technology of Food Preservation,* 4th Edition, AVI Publishing Co. Westport, Conn.

FAO. 1969. *Sun drying fruits and vegetables in Sudan.* Jackson, T.H. and Mohammed, B.B. FAO Report. Rome: FAO.

Feinberg, B. 1973. "Vegetables". In *Food Dehydration*, 2nd Edition, Vol. 2. W.B. Van Arsdel, M.J. Copley and A.I. Morgan, Jr.(Editors). AVI Publishing Co., Westport, Conn.

9

Processing for Sale

Interest in small scale food processing as a means of enterprise creation in developing countries has increased substantially in recent years, as a result of a number of factors. These include:

- greater promotion of private sector micro- and small-scale enterprises by national and international development agencies, to help increase incomes and employment opportunities in both rural and urban areas
- development of tastes for non-traditional foods, especially in major urban centres
- a desire by national governments to both increase export opportunities and foreign exchange earnings and to develop national manufacturing capacities
- increased food surpluses in some countries that require preservation and processing as a consequence of successful agricultural interventions or development programmes
- an increasing pride in locally produced products that can effectively compete with imported goods.

However, successful fruit and vegetable processing enterprises require more than the skills that are needed for home processing, described in Part 1. Whereas in home or village processing, the consumers and processors know each other and can give feedback on the quality or perceived value of a processed food, in formal food processing businesses the producers usually do not know who eats their food or what their customers think of the products. Similarly the customers have no direct link to the individuals who have produced their food and there are thus very limited opportunities for feedback between processors and consumers. As a result, processors have to learn new skills in developing attractive packaging and in marketing and selling if they are to successfully find customers.

Additionally, consumers expect to find the same food in every pack, having a uniform quality, every time that they buy a product. Producers must therefore control their process to produce uniformity and consistency in their products, and to do this they must learn new skills in quality assurance. There are also other problems that are associated with operating any enterprise that need to be addressed by aspiring entrepreneurs: for example legal aspects such as registration of the business, payment of taxes, employment conditions and business planning.

For those who set out to earn their main income from fruit and vegetable processing, there are special problems which make this type of business different from most others:

- many raw materials are highly perishable and will spoil quickly after harvest unless they are processed quickly
- many fruits and vegetables are also highly seasonal which means that the business can only operate for part of the year, unless crops are either part-processed for intermediate storage, or a succession of crops is processed throughout the year
- seasonably also affects the cashflow in an enterprise, as most raw materials have to be bought during a relatively short harvest period
- yields of fruit and vegetables are subject to considerable variation according to the weather, especially the rainfall patterns, and plant diseases. This can result in an unpredictable supply and large variations in prices for raw materials, which makes business planning more difficult
- some processed foods also have a seasonal demand which further complicates planning and cashflow of a business
- even after processing, some fruit and vegetable products have a limited shelf-life and distribution and sales methods must therefore be organized so that customers receive products in the required amounts before they spoil

- processing must be done to high standards of hygiene and production control to avoid the risk of harming or even killing customers by allowing contamination of the products by foreign materials or the growth of food poisoning micro-organisms.

Fruit and vegetable processing as a business is therefore a more complex activity than processing at home. Although it is true that most people can be trained to make high quality products, it is misleading to say that a person who is able to make a good product at home can then become a successful entrepreneur. Many (possibly most) small scale food processors start by working from home, but their ultimate success is dependent on a number of factors, including their planning and business skills, their creative flair to produce products that have a demand and are different from those of competitors and their determination to succeed.

Characteristically, successful entrepreneurs see food processing as their main source of income. They may take out a loan to buy specialized equipment or secure working capital and they develop business and marketing skills to expand and diversify their enterprise.

There are different definitions of micro- and small-scale enterprises and in this book, the following are used:

— A micro-enterprise is one in which the owner works each day to manage the business and produce food. The number of workers is less than 10, the total investment does not exceed $20,000 and annual sales are less than $12,000.

— A small scale enterprise is one in which the owners may work at the processing unit or employ a manager. The number of workers is less than 20, the total investment does not exceed $50,000 and annual sales are less than $25,000.

The different skills required by producers to successfully operate a small enterprise and earn a family income are described in this Part of the book. The availability of services (electricity, clean water, gas, servicing and maintenance skills and facilities etc.) obviously vary between different countries and even between regions of the same country. Similarly the amount of money available to invest and opportunities for obtaining credit also vary considerably. It is therefore not possible to describe precise conditions under which any business can be successful, but where appropriate, guidelines are given on the types of activity and investments that have made other small fruit and vegetable processing enterprises successful.

The criteria that need to be taken into account when deciding which technologies to use are complex and inter-related but are likely to include the following:

- technical effectiveness (will the equipment do the job required at the indicated scale of production)
- relative costs for both purchase and maintenance of equipment and any ancillary services required
- operating costs and overall financial profitability
- health and safety measures in the process
- conformity with existing administrative or production conditions
- training and skill levels required for operation and maintenance
- environmental impact, such as pollution of air or local waterways.

However, these factors are not simply a checklist and each will have a different weighting in different circumstances. There can be no simple solution to the difficult task of assessing all factors in a particular situation and making the 'best-fit' from the available solutions.

In every proposed development, the general principle when starting a business, having decided which product to make, is as follows: conduct market research and produce a draft feasibility study. If the business appears to be profitable, produce a business plan, obtain credit where necessary and register the business with taxation and local government authorities. Then establish production facilities, develop contracts or agreements with suppliers and customers and manage the routine production of the products. Continually review and update business plans to take account of the actions of competitors or changing markets.

Selecting Products and Production Methods

Fried Products

A small number of starchy fruits, including jakfruit, breadfruit and banana are fried and eaten as snackfoods. Heat during frying destroys enzymes and micro-organisms and if sufficient moisture is removed and the product is packaged, it can have a shelf life of several weeks. The most common example of this type of snack-food is banana chips and the process chart shows their production.

The stages in processing are described in the left-hand column and are split into those steps that are essential for all products and those that are only needed for some products.

Care is needed to control the temperature of the oil used for frying, not only for safety reasons, as very hot oil can splash onto operators when wet fruit is immersed, but also because of financial and quality considerations. When oil is heated too much, it exceeds its *smoke point* and a blue haze appears above the oil. This is a sign that the oil is breaking down chemically and it will then begin to get more viscous and develop an unpleasant flavour. The flavour is transferred to the product, making it unacceptable. When the oil thickens, more is retained on the product and there is a higher cost in buying more oil than is needed. Too much oil on the product also reduces its shelf life.

Bottled and Canned Products

Bottling and canning are essentially similar processes in that food is filled into a container and heated to destroy enzymes and micro-organisms. Fruits can be packed into jars with a hot, sugar syrup and vegetables can be packed into a hot brine. The filled jars are sealed and pasteurized so that an internal vacuum forms when they are cool. The sealed container then preserves the food by preventing re-contamination and excluding air and sometimes light. Preservation depends on an adequate heat treatment and an air-tight (or 'hermetic') seal.

There are three grades of syrup: a light syrup contains 200g sugar per litre, a medium syrup 400-600 g/l and a heavy syrup 800g/l. The concentration of salt in brine is usually 15 g/l. Acidic fruits require relatively mild heating conditions for pasteurization (e.g. 90-100°C for 10-20 minutes) to destroy yeasts and moulds, whereas less acidic vegetables require more severe heat sterilization to destroy food poisoning bacteria (e.g. 121°C for 15-40 minutes, depending on the size of the container). Fruits can also be part-processed and stored until required in sugar syrup or sodium metabisulphite solution (equivalent to 1000 ppm.) to allow production to take place for a larger part of the year.

It is not advisable for inexperienced small scale processors to bottle vegetables unless they are acidified, because of the risk of poisoning from inadequately processed foods. Vegetables can however be processed by pasteurization if the acidity is first adjusted using citric acid or vinegar.

Canning is not suitable for small scale processing for the following reasons: the time and temperature of canning are critically important and must be carefully controlled. If the cans are under-processed, there is a risk of serious food poisoning and even death from a type of micro-organism named *Clostridium botulinum*. If cans are over-processed, the vegetables lose much of their texture, colour, vitamins and flavour and are not saleable. The establishment of correct heating conditions depends on the type of food, the size and shape of the can and the initial level of contamination of the vegetables. This requires the skills of a qualified food technologist or microbiologist.

When foods are heated in sealed cans during the canning process, the temperature of sterilization is above 100°C and the pressure outside the can must equal that inside, to prevent the cans from exploding. This is achieved using high pressure steam and a strong vessel named a 'retort'. Both steam boiler and retort are expensive and likely to be beyond the means of a small scale processor. Additionally, compressed air is needed to maintain the pressure while cans are being cooled, which together with the necessary controllers, adds to the capital cost of equipment.

Even if cans are available in a particular country, they are usually more expensive than other forms of packaging. Different types of product also require a particular internal lacquer to prevent the metal from corroding when it is in contact with the fruits or vegetables and such lacquers may not always be available. In addition a 'seamer' is needed to seal the lid onto the can and regular checks and maintenance are necessary to ensure that the seam is properly formed. Failures in seams are one of the main causes of spoiled or dangerous canned foods.

It is therefore necessary to ensure that seamer operators are fully trained and experienced in adjusting the machines and a 'seam micrometer' is another necessary capital expense to be able to do this. In summary therefore, canning requires a considerable capital investment, trained and experienced staff, regular maintenance of relatively sophisticated equipment, a regular supply of the correct types of cans and a comparatively high operating expenditure.

Because of the more acidic nature of fruits, a lower processing temperature is adequate and this process is suitable for small scale operations. In all cases, a food technologist should be consulted to advise on process times and conditions for bottled products.

Dried Fruits and Vegetables

Drying removes most of the water from fruits and vegetables to extend their shelf life and to increase their convenience and value. The reduction in weight and bulk also makes transport cheaper and easier although many dried foods are fragile and require packing in boxes to prevent them from being crushed. Different categories of dried foods can be described as *high-volume, lower-value* crops such as staple cereals and *low-volume, higher-value* foods such as dried fruits, vegetables herbs and spices. This second category offers better opportunities for profitable production by small scale processors.

Air dried products are the most common type of dried fruit and vegetables and other more expensive methods, such as freeze drying, are not considered in this book. Some products may be blanched or sulphured/ sulphited to protect their natural colour and help preserve them. Crystallized fruits, peels for marmalade and cake production and osmotically dried fruits (known as 'osmasol' products when dried in a solar dryer) are fruit pieces that are soaked in hot concentrated sugar syrups to extract some of the water before air drying.

Some vegetables and a few fruits such as limes may also be salted before drying. In this case the high salt concentration preserves the food by both drawing out water by osmosis and by the anti-microbial properties of the salt. Salt tolerant micro-organisms begin to grow while the product is sun dried and these produce acids and characteristic flavours, High salt concentrations also prevent the action of some enzymes, which would cause a loss in quality of the dried food during storage. Vegetables must be washed to lower the salt concentration before they are eaten

Fruits and vegetables must be carefully selected before drying. If fruits in particular are over-ripe they are easily damaged and may be difficult to dry. If they are under-ripe, they have a poorer flavour, colour and appearance. Care and attention to hygiene are essential because any bacteria or moulds that contaminate vegetables before drying are likely to survive on the dried food. The temperature of drying is not high enough to kill them and when the food is re-hydrated, they can grow again and cause food poisoning.

Blanching

Blanching destroys enzymes and prevents changes in colour, flavour and texture during storage. However, by itself, it does not preserve the food and vegetables must therefore be further processed by drying to achieve a long shelf life. Vegetables are blanched by heating in hot water or steam for a short time and then cooled on trays. For production at a small scale, vegetables can be placed in a wire basket and immersed in boiling water.

In steam blanching, vegetables are placed in a strainer and this is then fitted over a pan of boiling water and covered with a lid to prevent the steam escaping. Steaming takes a few minutes longer than water blanching, but has the advantage of retaining more nutrients as they are not lost into the water.

There are optional chemical treatments that help to retain the colour and texture of some dried fruits and vegetables. For example, the bright green colour of leafy vegetables, peas etc. can be retained by adding sodium bicarbonate to blancher water and the texture of some vegetables, such as okra and green beans, can be maintained by blanching in a calcium chloride solution. Both chemicals are usually available from pharmacies in major towns.

Sulphuring and sulphiting

For most fruits, 350-400g sulphur are used per 100 kg fruit, burning for 1-3 hours. Sulphur dioxide prevents browning in foods such as apple, apricot and coconut, although it should not be used with red fruits as it bleaches the colour. Sulphuring (using sulphur dioxide gas) is achieved by exposing pieces of cut or shredded fruits to burning sulphur in a sulphuring cabinet. The amount of sulphur used and the time of exposure depend on the type of fruit, its moisture content and limits placed by law in some countries on the residual amounts of sulphur dioxide in the final product or by commercial limits set by importers. This should be checked with a local Bureau of Standards. There is an increasing consumer resistance to sulphited fruits in some industrialised countries and if the product is considered for export the local Export Development Board or import agents should be consulted.

In sulphiting, the sulphur dioxide is dissolved in water, rather than as the gas used in the sulphuring process. Sodium sulphite, sodium metabisulphite or potassium metabisulphite are made into solutions, either by adding one of them to the blanching water or more often, by soaking the food for 5-10 minutes in a sulphite dip. About two thirds of the weight of sodium metabisulphite is formed as sulphur dioxide when it is dissolved in water. For example, to form a 0.001% solution which is equivalent to 1000 ppm, 1.5g is dissolved in a litre of water to form 1g of sulphur dioxide per litre. At this concentration, sulphiting can also be used as a method of intermediate storage of fruits to spread production over several months throughout the year.

Syrup Pre-treatment

This method can be used to remove up to half of the water in fruit and is therefore a cheap way of increasing the production rate of a dryer or for part-processing fruits for intermediate storage so that production can be extended throughout the year. In general the method gives good retention of colour in the dried food and produces a sweeter, blander tasting product. However, acids are also removed from fruits during the process and the lower acidity of the product may allow mould growth if the food is not properly dried and packaged.

In a more complex method, fruit is first boiled in 20% syrup and then soaked overnight. The fruit is then strained from the syrup and transferred each day to 40% and 60% syrups in turn, with optional boiling for 10 minutes at each transfer. After soaking, the syrup is diluted to approximately half of the original concentration. Each day the most dilute syrup (10%) is used for other products and a new 60% syrup is made up. The advantages of this method include reuse of sugar syrups and a softer texture in the final product.

Types of Dryers

The higher value of dried fruit and vegetable products, compared for example to cereals crops, may justify the higher capital investment in a fuel-fired dryer or electric dryer and the extra operating costs for the fuel or electricity. These types of dryer allow higher drying rates and greater control over drying conditions than do solar or sun drying and they can therefore result in a higher product quality. However it is necessary to make a careful assessment of the expected increase in income from better quality products compared to the additional expense, to make sure that this type of dryer is cost-effective.

Sun drying is only possible in areas where, in an average year, the weather allows foods to be fully dried immediately after harvest. The main advantages of sun drying are the low capital and operating costs and the fact that little expertise is required.

The main problems with this method are as follows:

- contamination, theft or damage by birds, rats or insects
- slow or intermittent drying and no protection from rain or dew that wets the product, encourages mould growth and may result in a relatively high final moisture content
- low and variable quality of products due to over- or under-drying
- large areas of land needed for the shallow layers of food
- laborious because the crop must be turned, moved if it rains and animals must be kept away
- direct exposure to sunlight reduces the quality (colour and vitamin content) of some fruits and green vegetables.

The quality of sun dried foods can be improved by reducing the size of pieces to get faster drying and by drying on raised platforms, covered with cloth or netting to protect against animals and insects.

Solar drying has been studied in detail by scientists in many countries for many years, but it is not yet widely used commercially. The main applications to date are in Bangladesh for desiccated coconut, in Guatemala for herbal teas and in Uganda for dried fruits for export.

PACKAGING

If the climate is dry, it may not be necessary to package dried foods as they will not pick up moisture from the air. However a humid climate is likely to result in dried foods gaining moisture and going mouldy. The stability of dried foods depends not only on the humidity of the air at which a food neither gains nor loses weight, but also on the type of food. Different foods can be grouped according to their ability to absorb moisture from the air. The two groups are *hygroscopic*, which absorb moisture easily and *non-hygroscopic*, which do not absorb moisture. The classic example is salt and pepper, where salt is very hygroscopic and pepper is non-hygroscopic, but similar examples exist for fruit and vegetable products. This difference determines the packaging requirement for different fruit and vegetable products. The moisture content at which a food is stable is known as the*Equilibrium Moisture Content* and examples of this for different fruits and vegetables are shown in Table 8, together with the packaging requirement for different groups of foods.

Dried fruits and vegetables are usually packaged in one of the many different types of plastic film. The selection of the correct type of packaging material depends on a complex mix of considerations which include:

- the temperature and humidity of the air in which the product is stored
- the capacity of the product to pick up moisture from the air
- reactions within the product caused by air or sunlight during storage
- the expected shelf life
- marketing considerations
- cost and availability of different packaging materials.

In general, although thin polythene film is usually the cheapest and most widely available material, it is only suitable for storing dried fruits and vegetables for a short time before they pick up moisture, soften and go mouldy. Polypropylene has better barrier properties and therefore gives a longer shelf life, but it is usually more expensive and it may not be available in many countries. Other more complex films, such as laminated films made from polythene and aluminium foil, offer much better protection to dried foods, but are considerably more expensive and more difficult to find in developing countries.

Most dried foods also need a sturdy box or carton to both prevent crushing and to exclude light which causes loss of colour and development of off-flavours during storage.

From the table, it can be seen that some types of packages provide good protection against air and moisture pickup for example, whereas other protect against light, crushing, etc. It is therefore common for dried foods to be packed in airtight and moisture-proof bags, which are then placed in an outer container to protect against light, crushing, etc.

Chutneys, Pickles and Salted Vegetables

Chutneys

Chutneys are thick, jam-like mixtures made from a variety of fruits and vegetables, sugar, spices and sometimes vinegar. Any edible sour fruit can be used as a base for a chutney, to complement the sweet taste from the sugar. The high sugar content has a preservative effect and vinegar addition is not always necessary, depending on the natural acidity and maturity of the fruits that are used. Most products are boiled, which not only produces a caramelised syrup and alters the taste, colour and thickness, but also pasteurises the product and thus adds to the preservative action of the sugar and acids. Other products are allowed to ferment naturally and the acids produced by a mixture of bacteria preserve the product. Depending on the types of spices that are added, these may also have a preservative effect, in addition to their contribution to flavour. Fruits can also be sulphited as a method of intermediate storage to spread production over several months throughout the year.

Natural acids from the fruit, from vinegar or those produced by fermentation, together with the high sugar content, are used to preserve the chutney after a jar has been opened. A correct balance between the levels of sugar and acid is required to prevent mould growth and a *Preservation Index* can be used to calculate the amounts of ingredients to be added. Alternatively, when sugar is the main ingredient or the product is boiled, a refractometer can be used to check that the final sugar content of the syrup is 68-70%. Sugar is added before heating if a dark product is required or towards the end of boiling to produce a light coloured product.

The Preservation Index is a measure of the preserving power of combinations of acid and sugar (sugar is measured as 'total solids'). This is used to assess whether a chutney or pickle is safe from food spoilage and food poisoning micro-organisms. The value can be calculated as follows:

Pickles

Vegetables such as cucumber, cabbage, olive and onion are fermented by lactic acid bacteria which can grow in low concentrations of salt. The bacteria ferment sugars in the food to form lactic acid, which then prevents the growth of food poisoning bacteria and moulds or other spoilage micro-organisms. The amount of added salt controls the type and rate of the fermentation. If for example 2-5% salt is used, a natural sequence of different types of bacteria produce the lactic acid. If higher concentrations of salt (up to 16%) are used, a different product called 'salt stock' pickle is produced, which is preserved by the salt and not by fermentation. Fruits and vegetables can be preserved in this way as a method of intermediate storage to spread production over several months throughout the year. Sometimes, sugar is added to increase the rate of fermentation or to make the product sweeter. Alternatively, vegetables may be packed in vinegar (acetic acid), salt and sometimes sugar to produce a variety of pickled products. Because these vegetables are not fermented they have a different flavour and texture. They are usually pasteurised. Sweet pickles are made from single fruits or mixtures of fruits and vegetables. They are preserved by the combined action of lactic or acetic acid, sugar and in some cases added spices.

Salted Vegetables

Salted vegetables are made by building up alternate layers of chopped or shredded vegetable such as cabbage, with layers of salt in a sealed drum. The salt has two preservative actions: it draws out water from the vegetables by osmosis to form a concentrated brine in the base of the drum; and salt also has a direct anti-microbial action. The high levels of salt are reduced by washing the products before they are eaten.

Pickles which have an adequate Preservation Index do not need to be pasteurised. However as an additional measure to prevent spoilage, they can be pasteurised or the sugar/salt/vinegar mixture can be heated, added to the vegetables and the jars filled while product is still hot. In this way the hot product forms a partial vacuum in the jar when it cools and further aids preservation.

Glass jars are the most commonly used packaging material, but if a shorter shelf life is expected, pickles may also be packed in small quantities in polythene pouches and sealed with an electric heat sealer. To avoid seepage of product, which can damage paper labels and make the package unattractive, a double pouch can be used comprising an inner pack that contains the product and an outer pouch with a label between the two.

Pectin and Papain

Pectin

Pectin is a component of nearly all fruits and vegetables and can be extracted and used in food processing to form the characteristic gel in jams and marmalade. The richest sources of pectin are the peels of citrus fruits such as lime, lemon, orange and passion fruit or the pulp of apple (known as 'pomace' after juice has been extracted). Commercially, pectin is available as either a light brown powder or as a dark liquid concentrate. It is stable if stored in cool, dry place and it will lose only about 2% of its gelling power per year.

There are two main types of pectin:

1) high methoxyl (HM) pectins that form gels in high solds jams (above 55% solids) in a pH range of 2.0-3.5; and
2) low methoxyl (LM) pectins, which do not need sugar or acid to form a gel, but instead use calcium salts.

These form a gel with a wide range of solids (10-80%) within broader pH range of 2.5-6.5. They are used mainly for spreads or for gelling agents in milk products. However, there are a large number of different types of pectin within each group, such as 'rapid set' and 'slow set' pectins, that are made for different applications and it is necessary to specify carefully the type required when ordering from a supplier.

It is difficult to make specific grades of pectin or the powdered products at a small scale. Where attempts have been made to set up small commercial pectin units, the process has not been financially viable due to the expense of recovering solvents that are used to precipitate the pectin before drying. However, it is possible for jam-makers to increase the amount of crude (unrefined) pectin in their products by extracting it with water without using solvents. This is useful if the producer is making jam from fruits such as melon that are naturally low in pectin. When using pectin to make set preserves, such as jams, jellies etc., it is important that the pH is within the range of 3.0-3.5 and that they are boiled to at least 68% solids content. Powdered pectin is added to fruit pulp at 3-6g per kg of final product, but it should first be mixed with about five times its weight of sugar to prevent lumps forming when it is added to the pulp or juice

Papain

Papain is an enzyme that is found in the skin of unripe papaya fruits. It breaks down proteins and finds widespread applications in industrialised countries for meat tenderising and in brewing. Advances in biotechnology during the last twenty years resulted in a synthetic papain that was cheaper than the natural enzyme and as a result there was a decline in the market for natural papain. However, in recent years, there has again been increasing interest in the natural product and it can now achieve a higher value than before.

For profitable production, it is usually necessary to plant papaya trees in orchards, rather than collecting it from widely scattered trees which increases collection costs. Crude (unrefined) papain is normally produced because the refining technology is too expensive for most small-scale entrepreneurs. The process of papain extraction involves making shallow cuts in the skin of unripe fruits while they are still on the tree. The skin then oozes a white, sticky liquid which is collected by scraping the fruit each day for several weeks. This 'latex' is then spread out in shallow trays and sun-dried until it becomes brittle. The crude papain is packaged and exported to refiners. Because papain is an enzyme that attacks proteins, the crude latex should be handled with gloves to avoid damage to the operator's skin.

Sauces

Sauces are thick viscous liquids, made from pulped fruit and/or vegetables with the addition of salt, sugar, spices and vinegar. They are pasteurised to give the required shelf life, but the basic principle of preservation is the use of vinegar, which inhibits the growth of spoilage and food poisoning micro-organisms. Other ingredients such as salt and sugar contribute to the preservative effect and the correct Preservation Indexensures that the product does not spoil after opening and can be used a little at a time. Some may contain a preservative such as sodium benzoate, but this is not necessary if an adequate Preservation Index is achieved. Sauces can be made from almost any combination of fruit or vegetables, but in practice the market in many countries is dominated by tomato sauce, chilli sauce and to a lesser extent, mixed fruit sauces such as 'Worcester' sauce, which contains apples and dates in addition to tomatoes. Depending on the scale of production, pulping and sieving out seeds and skins can be done by hand or using special pulper-finisher machines.

Similarly, at a small scale, sauces can be made using simple open boiling pans, provided that care is taken to heat slowly with constant stirring to avoid localised burning of the product, especially at the end of heating. At a larger scale, processing is done using steam heated, stainless steel 'double jacketed' pans.

Juices

The beverage market is divided into products that are intended to quench thirst, such as juices, nectars and carbonated drinks and those that are drunk on special occasions, such as wines and spirits, where religion or custom permit. Competition from large scale manufacturers is among the fiercest in beverage manufacture and considerable amounts of money are spent on advertising, packaging and sophisticated distribution systems in this sub-sector. Thus beverage manufacture is one of the most difficult for small scale producers to become established and successful. However there is a growing trend in urban centres of some developing countries towards increased juice consumption and this market may become larger in future years.

Acidity

pH paper is useful for checking the acidity of some less acidic juices (e.g. melon juice at pH 4.0-5.0). The optimum range is 3.0-4.0 and it may be necessary either to blend juices or adjust the pH with citric acid, to prevent spoilage of less acidic juices.

Pulping

Soft fruits such as berries, passion fruit and papaya can be pulped by hand, whereas harder and more fibrous fruits such as pineapple and mango require mechanised pulping. Manually operated pulpers are available for places where there is no electricity and there are also a wide range of powered pulpers and blenders. At higher production rates, pulper-finishers brush fruit through a sieve and separate seeds, skins etc. from the pulp. Juice can also be extracted from fruit using a fruit press or fruit mill or by steaming the fruit. Citrus fruit juices are extracted by reaming the fruit and again comparatively simple equipment is available for this purpose.

Pasteurisation

Drinks need to be pasteurised if they are to have a shelf life of more than a few days. The time and temperature required for pasteurisation depends on the type of product and the bottle size, but is typically 10-20 minutes at 80-90°C. They can either be pasteurised and then hot filled into pre-sterilised bottles, or cold filled and then pasteurised in sealed bottles using a large pan of simmering water with the water level around the shoulder of the bottle. Because the pasteurisation temperature does not exceed 100°C, there is little risk of the bottles bursting. However, as with all glass containers, the temperature difference between the glass and the product or hot water should not exceed 20°C. A water cooler can be constructed to speed up the rate of cooling of filled containers. Hot bottles and cool water pass in a counter-current way through a trough to minimise 'heat shock' to the containers. In order to prevent contamination by dust, insects etc., all bottles must be cleaned properly using hand-held bottle brushes or mechanised brush cleaners. If bottles are re-used, they must be thoroughly washed with detergent to remove any residual material that may have been stored in the bottle and then sterilised by heating to boiling in a water-bath for at least 10 minutes.

Filling

Hand filling from a jug is often too slow for the required throughput and fillers can be constructed by fitting one or more taps to the base of a stainless steel or food grade plastic bucket. More sophisticated fillers measure and control the volume of liquid filled into each bottle.

Waxed cartons have become popular because of their lower cost compared to bottles, together with savings in time and money for collection and washing of re-useable bottles. However, the cost of equipment to form and seal the cartons is too high for most small scale producers and cheaper alternatives including plastic pots with sealed foil lids are becoming more popular in some countries.

Squashes, Cordials and Syrups

Squashes and Cordials

These are drinks that are diluted to taste with water and are thus used a little at a time. The container must therefore be re-closeable and these products may contain a preservative, usually sodium benzoate, to prevent spoilage after opening. Squashes are made from at least 30% fruit juice mixed with sugar syrup. Cordials are simply crystal-clear squashes. Although food dyes are used by some processors, these are not necessary for most products. Regulations on the composition of squashes are in force in some countries.

Syrups

Syrups are filtered juices that are concentrated by boiling until the sugar content reaches 50-70%. The heat and high solids content preserves the syrup and it is used in place of sugar or honey. Syrups can be made from a wide range of fruits, but the most common type is made from grapes.

PRESERVES

Jams, Jellies and Marmalades

Jam is a solid gel made from fruit pulp or juice from a single fruit or from a combination of fruits. The composition is controlled by law in some countries. In mixed fruit jams, the first named fruit should be at least 50% of the total fruit content. The sugar content is normally 68-72%, which will prevent mould growth after opening the jar.

Jellies are crystal-clear jams that are produced using filtered juice instead of fruit pulp. Marmalades are produced mainly from clear citrus juices (including lime, orange, grapefruit, lemon and orange) and have fine shreds of peel suspended in the gel. Ginger may also be used alone or in combination with the citrus fruits. The fruit content should not be less than 20% citrus fruit and the sugar content is similar to jam. The correct combination of acid, sugar and pectin is needed to achieve the required gel structure and rapid boiling is necessary to remove water quickly, to concentrate the mixture before it darkens and loses its ability to form a gel.

Pastes and Purees

In principle, pastes and purees can be made from any fruit or vegetable. The most common types are tomato and garlic, which are widely used in cooking. These products can be made at a small scale by carefully evaporating water to concentrate the pulp, with constant stirring to prevent darkening or localised burning. The concentration of solids in the paste is normally around 36%. The high solids content and natural acidity are sufficient to preserve the product for several days but pasteurisation in bottles or cans is needed for a longer shelf life. In some preparations, sugar, salt and vinegar are added to assist in preservation.

Fruit Cheeses

Fruit cheeses are fruit pulps that are boiled until they have a final sugar content of 75-85%. They set as a solid block without the need for added pectin, but they do not have the same gel structure as jams and marmalades. They can be cut into bars or cubes to eat directly or they can be used in small pieces in confectionery or baked goods.

Boiling

Boiling is done using a stainless steel pan. If other materials are used there is the risk that fruit acids will react with the pan and cause off flavours to develop. At higher production rates, a steam jacketed pan is preferable as it gives more even and faster heating. If whole fruit is used, there are two heating stages in jam manufacture: initially fruit is heated slowly to soften the flesh and to extract pectin; then the mixture is boiled rapidly until the sugar content reaches 68-72%. This change in heat output requires a sufficiently large and easily controllable heat source. There are three ways to test for the correct point to stop boiling: using a hand-held refractometer reading 68-70% sugar, a sugar thermometer reading 104-105°C or by placing a drop of the product in cold water to see if it sets.

An alternative method to concentration by boiling, when making products such as tomato paste, is to hang the pulp in a sterilised cotton sack for an hour. During this time the thin watery juice leaks out and the pulp loses half its weight. Then 2.5% salt is mixed into the concentrate and it is re-hung for a further hour, during which time the weight falls to one third of the original. The product can then be packaged and pasteurised or further concentrated. This method is reported to produce a product which has a natural flavour using considerably less fuel than concentration by boiling.

Preserves should be hot filled into new or re-used glass jars, which are then sealed with a new lid. In some countries, plastic containers have been introduced but they are not easy to hot fill (they melt) and the seals are often inadequate, causing product leakage and insect infestation problems as well as a shorter shelf life.

The temperature of filling should be around 85°C. If it is too high, steam condenses on the inside of the lid and water falls onto the surface of the preserve. This dilutes the sugar at the surface and results in mould growth.

If the temperature is too low, the preserve thickens and is difficult to fill into containers. Filling can be done using jugs and funnels, but for higher production rates, small hand-operated or semi-automatic piston fillers are available. In all cases the jars should be filled to approximately 9/10ths full, to assist in the formation of a partial vacuum in the space above the product as it cools. This can be checked using a 'headspace gauge'. The jars are kept upright until the gel has formed during cooling.

Wines, Vinegars and Spirits

Wines are produced by fermentation in which the sugars in the fruit juice/

pulp plus added sugar are converted into alcohol and carbon dioxide by varieties of the yeast *Saccharomyces cerevisiae*, named 'wine yeasts'. Producers need to select one variety that works well in their process and then continue to use it to produce a consistent product. Wines are preserved by the raised levels of alcohol and their natural acidity. Almost any fruit can be used to make wine, but the most popular in many developing countries are pineapple, papaya, grape, passionfruit, banana, melon and strawberry (or strawberry flavoured wine). Typically, the alcohol content of wine is 6-12% and in 'fortified' wines, such as sherry, ginger wine etc., it is usually 15-20%. The main problems are concerned with adequately cleaned fermentation vessels, to prevent contamination by other micro-organisms that spoil the wine and adequate sedimentation or filtration to produce a crystal-clear product. A special licence to sell alcohol is needed in most countries.

Vinegar is produced by a second fermentation in which *acetic acid bacteria (Acetobacter species)* convert the alcohol in wine to produce acetic acid, (spoiled wine is often due to the unwanted presence of acetic acid bacteria). Whereas in wine production, air is excluded from the vessel by an air lock, in this second fermentation it is important for the bacteria to be in contact with air as much as possible. This is done traditionally by allowing the wine to trickle over a stack of wood shavings or twigs that are held in an open framework, named a 'generator' which allows free circulation of air. The wood becomes covered in the bacteria and the process either operates continuously, re-circulating the wine a number of times until the acetic acid levels rise sufficiently, or part of a batch is retained to inoculate the next batch. Typically, vinegar contains 6-10% acetic acid, which preserves the product for many months/years provided that it is sealed in an airtight container to prevent the acetic acid from evaporating. More modem vinegar fermenters pump air into the wine and are sufficiently expensive to be beyond the reach of most small scale producers.

Alcohol has a lower boiling point than water and distillation (vaporising the alcohol and then condensing it) is used to concentrate the alcohol in spirit drinks. Distillation is carried out in a still which is a clean drum, fitted with a safety valve and a pipe to carry away the vapour. Wine or other alcoholic liquor is placed inside the drum and heated. The alcohol vapour is passed through cooled air or cool water and the distillate condenses and is collected.

There are a large number of traditional methods for distillation. One such method in East Africa involves filling fruit wine or beer into an oil drum and placing it at a shallow horizontal angle over a fire (this angle prevents any liquor from being sucked over with the distillate during boiling). A hollow bamboo pipe, approximately 10 cm in diameter, is fitted to the opening of the drum and sealed with a pulp made from the inner layers of a banana trunk. Five hollow bamboo pipes, approximately 2 cm in diameter, are fitted vertically down from the larger pipe and sealed in a similar way. Each pipe leads to a baked clay pot of approximately 20 litres capacity, submerged in a pit of slow-running water, diverted from a stream. The pots are held below the surface of the water by wood branches that are bent over them and fixed into the soil at each end.

Distillation is controlled by sampling each pot at intervals to determine its alcohol strength and adjusting the rate of heating by adding or removing firewood. As the strength declines with continued distillation, the end point is determined by the desired alcohol strength in the product, usually 35-40%. Distillation is continued for 3-4 hours and 40 litres of spirit are filled from the pots into clean bottles and sealed with a stopper made from banana or papyrus leaves. The spirit is preserved by the high alcohol content and has a shelf life of several years.

Modem stills are made entirely of copper or stainless steel and larger ones are fitted with thermostatically controlled heaters. These are too expensive for most small scale producers, but small copper stills for making essential oils have been used for making small amounts of alcohol.

Sugar may be added to fruit juice to raise the level to about 20° Brix and after inoculation with about 3% yeast, it is held in a fermentation vessel, made from food grade plastic or glass, for about ten days. It is important to keep the vessel closed to prevent bacteria and mould from infecting the batch. After ten days the fermenting wine is 'racked' (filtered) by passing it through a muslin or nylon straining cloth into narrow necked fermentation vessels, plugged with cotton wool or fitted with a air lock.

The fermentation is then continued for between three weeks to three months, depending on the temperature and the strain of yeast being used. The end of the fermentation is seen when there are no more bubbles rising to the surface.

It is then siphoned into clean containers and allowed to clear and mature before it is siphoned into bottles and sealed with sterilised cork stoppers or roll on pilfer-proof (ROPP) screw caps. Some fruit wines, such as pineapple, are difficult to clear because of natural gums in the fruit. These can be removed by heating the juice and allowing the gums to precipitate out before the juice is fermented. This can be a particular problem when waste fruit from other processes (e.g. drying or bottling) is used for wine making. Great care should be taken to ensure that only the fermented liquor is used for distillation. Any suspect liquor or other materials should not be distilled because of the risk that they may contain methanol. This is a type of alcohol which if consumed causes blindness and in sufficient amounts, death.

References

Fellows, P.J., 1993, *Food Processing Technology*, Woodhead Publishing, Cambridge, UK

Central Food Technology Research Institute, 1977, *Home Scale Processing and Preservation of Fruits and Vegetables*,Mysore, India.

Ihekoronye, A.I., and Ngoddy, P.O., 1985, *Integrated Food Science and Technology for the Tropics,* Macmillan Press Ltd., London, UK.

Tindall, H.D., 1983, *Vegetables in the Tropics*, MacMillan Press Ltd., London.

MacDonald, I. and Low, J., 1984, *Fruit and Vegetables*, IT Publications, London, UK.

10

Quality Control

The international trade in processed fruits and vegetables is very large with an ever increasing number of different types being processed and exported. Whereas once, processing was limited to mostly temperate climate fruits and vegetables, the change has now broadened to include tropical and subtropical types.

The reasons are twofold. Firstly, consumers' dietary habits have become more diverse so that, for example people living in North America may very well like fruit and vegetables grown in Africa or Asia. Secondly, processing techniques, whether they be for canning, freezing or drying, have been improved to an extent where final product is palatable, nutritious and of long and reliable shelf life.

Many developing countries have taken advantage of the continuing worldwide demand for processed fruits and vegetables and earned valuable foreign exchange from exports of products to profitable markets.

The export quality control and inspection of processed fruits and vegetables is directed at ensuring that the final products:

- have been processed in a registered export establishment that is constructed, equipped and operated in an hygienic and efficient manner;
- conform to the requirements of the export regulations for processed fruits and vegetables, and those of the importing country, in respect of such things as quality grades, defects, ingredients, packaging materials, styles, additives, contaminants, fill of container, drained weight; and, conform to labelling requirements.

Inspection and Certification Procedures

In most countries, in processing fruits and vegetables for export, it is not customary to apply continuous inspection as it is in the case of meat. Few, if any, importing countries require it, and the nature of the products themselves is such that only part time check inspection is required during processing together with statistically based inspection, including sampling and analysis, of final product.

However, in circumstances where an establishment is processing export product for the first time, it can be argued that there is an advantage in adopting continuous inspection until the operation is satisfactorily established.

In any event, inspection of raw materials should be carried out at the commencement of each processing run to ensure that only sound fruit or vegetables of sufficient maturity (degree of ripeness) are used for processing. Sampling checks of raw materials should be carried out as frequently as the inspector thinks necessary.

The inspector must ensure that adequate hygiene practices are followed during the processing of the product. For example, in the case of canned and frozen products and other processing methods, raw materials should be washed absolutely clean so that fruit and vegetables entering the processing line are free from dirt, superficial residues of agricultural chemicals, insects and extraneous plant material.

In the case of dried product, especially where the raw material is sun-dried on drying greens or racks, care must be taken to minimize contamination by bird and animal droppings, dust and extraneous plant material. It is often necessary to wash the dried product to ensure cleanliness of the final product.

In the case of canning and freezing, the inspector must obtain full details of the processing programme for at least the following day from management, so that an adequate inspection programme can be scheduled.

In much the same way as for fresh fruit and vegetables, the inspector must also be aware of the pesticides and other chemicals used in the production of the raw materials. Necessary laboratory analyses can then be arranged to ensure residue levels in the final product do not exceed tolerances adopted by importing countries.

At the commencement of and during processing, the inspector should pay attention to the state of raw materials, the preparation of raw materials for processing (peeling, slicing, dicing, blanching, etc.), preparation and density of packing medium (sugar syrup, salt brine, etc.), the state of cans or containers to be used (cleanliness and strength), the cooking or freezing process (time/temperature relationship), can filling and closure and can/container storage.

After processing, the inspector should check the final product to ensure the drained or thawed weight, the vacuum and headspace, packing medium strength and that can/container conditions are satisfactory. Statistically based sampling plans should be adopted for the examination of final product to ensure it meets the requirements of the export regulations.

The labelling applied to cans/containers should also be checked to ensure both their correctness and compliance with the export regulations and the requirements of those countries in which the product is to be marketed.

Cans should also be examined to make sure that the correct embossing relating to the product, its date of production and the registered number of the export establishment has been applied.

Each establishment registered for the export of processed fruit and vegetables or for canned or frozen foods should have its own quality laboratory sufficiently equipped and staffed to carry out physical, chemical and microbiological examinations of the goods.

Inspectors should have access to the laboratory facilities and the establishment's quality control records as and when required. Independent laboratory examination of product should be made by the agency having responsibility for export on the basis of a statistically developed sampling plan.

In those countries where fruit and vegetable production is a seasonal event, processing for export generally takes place at the time of peak production and then declines, often to a halt, as the supply of raw materials declines. As a result, most export establishments produce at their peak of production far more product than they export at that time.

Therefore, most manufacturers find it necessary to store product for considerable periods before it is exported. Thus, proper storage is essential if the product is to retain its quality and cans remain untarnished. Inspectors should regularly inspect storage facilities, noting their conditions and that of the stored product, looking for signs of deterioration such as pest infestation and rusting of cans.

Prior to export, the exporter should be required to notify the export quality and inspection agency of his intention to export in accordance with the provisions of the export processed fruits and vegetables regulations and on the prescribed "Notice of Intention to Export" form.

The notice should be submitted in sufficient time before the shipment date to enable the product to be inspected satisfactorily; the intensity of inspection depending on the original state of the product, the conditions under which it has been stored and the length of storage. When product is approved, the agency will issue the exporter an "Export Permit" authorizing Custom's clearance of the product.

Labelling

Customers and consumers expect the labelling on food to be a true description of what they are buying.

Misleading or fraudulent labelling is an unfair trade practice that cannot be tolerated. Most countries now have labelling laws stipulating how foods are to be labelled and what information labels must contain. Most, if not all of those laws have in common requirement that the label should bear:

- a statement of identity and a true, as distinct from misleading, description of the product;
- a declaration of net contents (weight or number of pieces);
- the name and address of the manufacturer, packer, distributor or consignee, and
- a list of ingredients (in descending order of volume or weight).

In addition, labels may also be required to include, amongst other things, the country of origin, date of manufacture or packing, a use-by or expiry date, nutritional qualities or values of the food, storage directions, a quality grade and directions for preparing the food.

More frequently than is often realized, consignments of food exports arriving on foreign markets are not permitted entry because the labelling does not comply with the mandatory requirements of the importing country.

This sometimes results in consignments being rejected, but more often in them being withheld from entry until the labelling is corrected or new labelling applied. In either case, trade is interrupted and the cost involved may make sales unprofitable. It is essential therefore, that exporters be familiar with the food labelling requirements of importing countries.

Export Quality Control and Inspection Systems for Foods

With the advent and development of a food consciousness amongst consumers, stimulated by the work of the Joint FAO/WHO Codex Alimentarius Commission through its elaboration of food standards, codes of hygienic practice and the Code of Ethics for International Trade in Food, an increasing number of countries have adopted sophisticated food laws and established food control agencies, some with the aid of FAO.

Consequently, those countries no longer accept products on trust that they are satisfactory, but instead, demand that food imports meet the requirements of their food laws and pass inspection by their control agencies. Moreover, many of them require exporting countries to certify that products comply with their national legislation and some also require additional special declarations.

As a result of these developments the emphasis of activity of Export Quality Control and Inspection Systems has changed. Although most of them still establish their own standards of quality control and adopt standards for foods for export, most of their effort and resources are now directed at ensuring that foods for export meet the mandatory requirements of importing countries and providing the necessary associated certification. To do otherwise is to invite either the detention or, at worst, rejection of product at point of entry.

Detentions and Rejections

Food exporting countries can no longer assume that there is a good chance that products not complying with the requirements of importing countries will escape the inspection at the point of entry.

Details of foods imports released by the United States Food and Drug Administration (FDA) indicate that significant quantities of product are at least detained, and at worst rejected, because they fail to meet U.S. food laws.

Reasons given for the detentions include:

- non compliance with labelling requirements;

- decomposition;
- insect and animal filth and damage;
- use of prohibited additives;
- heavy metal contamination;
- excessive levels of pesticide residues;
- excessive levels of mycotoxin;
- mould infestation;
- microbiological contamination;
- swollen and otherwise faulty cans.

The message for food exporting countries is quite clear - ensure your products comply with the mandatory requirements of importing countries or run the very real risk of having them rejected at considerable financial loss to the exporter and the country and resulting in damage to the commercial reputation of both.

While the foregoing relates to the U.S.A. experience, because it is the only country that currently publishes data about detentions and rejections of food imports it can be assumed that record more or less reflects the experience of other food importing countries. It might well be asked why such significantly high levels of detentions and rejections of food imports take place.

Undoubtedly the reasons are many and varied. However, the evidence shows that the most important reasons include:

- the inability of some export food industries, especially in developing countries, to handle, process, package and transport products to meet the mandatory requirements of importing countries;
- lack of awareness by food exporting countries of the mandatory requirements of importing countries, including certification;
- lack of adequate export control programmes and related agencies in food exporting countries, preventing them from exercising the necessary product surveillance and giving reliable and credible certification, and
- a lack of communication, between food control authorities and agencies in exporting and importing countries.

All four can be remedied by governments if they possess sufficient political will and take the necessary steps to do so.

Good Manufacturing Practices

Personnel

Disease control

Any person who has an illness, open lesions, including boils, sores, infected wounds, or any other abnormal source of microbial contamination must not work in any operation (in a food processing centre) which could result in the food, food-contact surface, or food packaging materials becoming contaminated.

Cleanliness

The following applies to people who work in direct contact with food preparation, food ingredients or surfaces of equipment or utensils that will contact food: they must wear clean outer garments, maintain a high degree of personal cleanliness and conform to hygienic practices while on duty; they must wash their hands thoroughly and, if they are working at a job where it is necessary, they must also sanitize their hands before starting work, after each absence from the workstation and at any other time when the hands have become soiled or contaminated; they must also remove all unsecured jewelry. People who are actually handling food, should remove any jewelry that cannot be properly sanitized from their hands; it is necessary to wear effective hair restraints, such as hairnets, caps, headbands or beard covers; operators must not store clothing or other personal belongings in food processing areas. Also, eating food, drinking beverage or using tobacco (in any form) must not be allowed in food processing area; all necessary steps have to be taken by supervisors to prevent operators from contaminating foods with microorganisms or foreign substances such as perspiration, hair, cosmetics, tobacco, chemicals and medicants.

Education and training

Persons who are monitoring the sanitation programs must have the education and/or experience to demonstrate that they are qualified. Food handlers and supervisors should receive training that will make them aware of the danger of poor personal hygiene and unsanitary work habits.

Supervision

Someone must be assigned the responsibility that all personnel will comply with all the requirements of these GMP's.

Plants and grounds

Grounds around a food processing centre which are under the control of this centre must be free from conditions such as: improperly stored equipment; litter, waste or refuse; uncut weeds or grass close to buildings; excessively dusty roads, yards or parking lots; inadequately drained areas - potential foot-borne filth or breeding places for insects or microorganisms; inadequately operated systems for waste treatment and disposal. 10.2.2.2. Plant construction and design shall: provide enough space for sanitary arrangement of equipment and storage of materials; floors, walls and ceilings must be constructed so that they are cleanable and must be kept clean and in good repair; separate by partition, location, time and other means, any operations that may cause cross-contamination of food products with undesirable microorganisms, chemicals, filth or other extraneous material; provide effective screening or other protection to keep out birds, animals and vermin such as insects and rodents. provide adequate ventilation to prevent contamination of foods with odours, noxious fumes or vapours (including steam); light bulbs, skylights or any other glass must be of the safety type or protected so that glass contamination cannot occur in case of breakage.

SANITARY OPERATIONS

General Maintenance

The plant and all fixtures must be kept in good repair and be maintained in a sanitary condition. Cleaning operations must be conducted in a manner that will minimize the possibility of contaminating foods or equipment surfaces that contact food.

Pest Control

- No animals or birds are allowed anywhere in the plant
- Programs must be in effect to prevent contamination by animals, birds and pests, such as rodents and insects;
- Insecticides and rodenticides may be used as long as they are used properly (according to label instructions);
- These pesticides must not contaminate food or packaging materials with illegal residues;

Sanitation of Equipment and Utensils

- Utensils and equipment surfaces that are in contact with food must be cleaned as often as necessary to prevent food contamination;
- Equipment surfaces that are not in contact with food should be cleaned as frequently as necessary to minimize accumulation of dust, dirt, food particles, etc.
- Single-service articles such as disposable utensils, paper cups, paper towels, etc., should be:
- Handled, dispensed, used and disposed of in a manner that prevents contamination of food or equipment;
- Where there is the possibility of introducing undesirable microorganisms into food, all utensils and equipment surfaces that contact food must be cleaned and sanitized before use and following any interruption during which they may have become contaminated;
- When utensils or equipment are used in a continuous production operation, they must be cleaned and sanitized on a predetermined schedule;
- Any facility, procedure, machine or device may be used for cleaning and sanitizing, as long as it has been established that the procedure will do the job effectively.

Storage and Handling of Clean Portable Equipment and Utensils

a) This refers to portable equipment or utensils which have surfaces that will contact foods;

b) When such equipment or utensils have been cleaned and sanitized, they should be stored in a manner that will protect the food contact surfaces from splash, dust and other contamination.

Sanitary Facilities and Controls

Water supply

Any water that comes into contact food or processing equipment must be safe and of adequate sanitary quality.

Sewage disposal

Must flow into an adequate sewage system or disposed of through other adequate means.

Plumbing

Must be of adequate size and design to:

a. Supply enough water to areas in the plant where it is needed;

b. Properly convey sewage or disposable liquid waste from the plant;

c. Not create a source of contamination or unsanitary condition;

d. Provide adequate floor drainage where hosing-type cleaning is done or where operations discharge water or liquid waste onto the floor;

e. Insure that there is no backflow from cross-connection between piping systems that discharge waste water or sewage, and those that carry water for food or food manufacturing.

Toilet facilities

a. Toilets and hand-washing facilities must be provided inside the fruit and vegetable processing centres;

b. Toilet tissue must be provided;

c. Toilets must be kept sanitary and in good repair;

d. Toilet rooms must have self-closing doors;

e. Toilet rooms must not open directly into areas where food is exposed unless steps have been taken to prevent airborne contamination (example: double doors, positive airflow, etc.);

f. Signs must be posted that direct employees to wash their hands with soap or detergent after using the toilet.

Hand-washing facilities

a) Adequate and convenient hand-washing and, if necessary, hand-sanitizing facilities must be provided anywhere in the plant where the nature of employees jobs requires that they wash, sanitize and dry their hands;

b) These hand-washing facilities must provide:

- Running water at a suitable temperature;
- Effective hand-cleaning and hand-sanitizing preparations;
- Clean towel service or suitable drying devices;
- Easily cleanable waste receptacle;
- Water control valves designed and constructed to protect against recontamination of clean, sanitized hands;

- Signs directing employees handling unprotected food to wash and, if appropriate, sanitize theirs hands before starting work, after each absence from the workstation, and any other time when the hands have become soiled or contaminated.

Rubbish and offal disposal must be handled in such a manner that they do not serve to attract or harbour pests or create contaminating conditions.

Equipment and Utensils

a. Equipment and utensils must be designed and constructed so that they are adequately cleanable and will not adulterate food with lubricants, fuel, metal fragments, contaminated water, etc.

b. Equipment should be installed so that it, and the area around it, can be cleaned;

c. Food contact surfaces shall be made of nontoxic materials and must be corrosion-resistant;

d. Seams on food contact surfaces shall be smoothly bonded, or maintained in order to minimize the accumulation of food particles, dirt and organic matter;

f. Holding, conveying and manufacturing systems, including gravimetric, pneumatic, closed and automated systems, shall be maintained in a sanitary condition;

g. Each freezer and cold storage compartment shall have an indicating thermometer, temperature measuring or recording device, and should have an automatic control for regulating temperature, or an automatic alarm system to indicate a significant temperature change;

h. Instruments and controls used for measuring, regulating or recording temperatures, pH, acidity, water activity, etc. shall be adequate in number, accurate and maintained.

Processes and Controls

There must be an individual who is responsible for supervising the overall sanitation of the plant.

Raw Materials and Ingredients

a. Must be inspected and sorted to insure that they are clean, wholesome and fit for processing into human food;

b. Must be stored under conditions that will protect against contamination and minimize deterioration;

c. Must be washed or cleaned to remove soil and other contamination:
   - Water used for washing, rising or conveying food products must be of sanitary quality;
   - Water must not be reused for washing, rinsing or conveying if contamination of food may result;
   - Containers and carriers (such as trucks or railcars) should be inspected to assure that their condition has not contaminated raw ingredients;

d. Raw materials shall not contain levels of microorganisms that may produce food poisoning or other disease, or they shall be pasteurized or otherwise treated during manufacturing operations so that the product will not be adulterated;

e. Materials susceptible to contamination with natural toxins, e.g., aflatoxin, shall comply with national and international official levels before they are incorporated into the finished food;

f. Materials susceptible to contamination with pests, undesirable microorganisms, or extraneous material, shall comply with national and international regulations, guidelines and defect action levels;

g. Materials shall be stored in containers, and under conditions which protect against contamination;

h. Frozen materials shall be kept frozen. If thawing is required prior to use, it shall be done in a manner that prevents contamination.

Manufacturing Operations

a. Food processing equipment must be keep in a sanitary condition through frequent cleaning and, when necessary, sanitizing. If necessary, such equipment must be taken apart for thorough cleaning.

b. It is necessary to process, package and store food under conditions that will minimize the potential for undesirable microbiological growth, toxin formation, deterioration or contamination. To accomplish this may require careful monitoring of such factors as time, temperature, humidity, pressure, flow rate, etc. The object is to assure that mechanical breakdowns, time delays, temperature fluctuations or other factors do not allow the foods to decompose or become contaminated.

c. Food shall be held under conditions that prevent the growth of undesirable microorganisms as follows:
   - Refrigerated foods shall be maintained at 45° F or below;
   - Frozen foods shall be maintained in a frozen state;
   - Acid or acidified foods to be held in hermetically sealed containers at ambient temperatures shall be heat-treated to destroy mesophyllic microorganisms;

d. Measures such as sterilizing, irradiating, pasteurizing, etc., shall be adequate to destroy or prevent the growth of undesirable microorganisms;

e. Work-in-process shall be protected against contamination;

f. Finished food shall be protected from contamination;

g. Equipment, containers and utensils shall be constructed, handled and maintained to protect against contamination;

h. Measures, e.g., sieves, traps, metal detectors, shall be used to protect against the inclusion of metal or other extraneous material in food;

i. Food or materials that are adulterated shall be disposed of in a manner that prevents other food from being contaminated;

j. Mechanical manufacturing steps such as washing, peeling, etc., shall be performed to protect against contamination by providing adequate protection from contaminants that may drip, drain or be drawn into the food, by adequately cleaning and sanitizing all food-contact surfaces and by using time and temperature controls at and between each manufacturing step;

k. Heat-blanching should be done by heating the food to the required temperature, holding it at this temperature for the required time, and then either rapidly cooling the food or passing it to the next manufacturing step without delay;

l. Filling, assembling, packaging, and other operations shall be performed in such a way that the food is protected against contamination by:
   - Use of a quality control operation in which the Critical Control Points are identified and controlled during manufacturing;
   - Adequate cleaning and sanitizing of all food-contact surfaces and food containers;

- Using materials for food containers and food-packaging materials that are safe and suitable;
- Providing physical protection from contamination, particularly airborne contamination;
- Using sanitary handling procedures.

m. Food such as, but not limited to, dry mixes, nuts, intermediate moisture food, and dehydrated food, that relies on the control of aw for preventing the growth of undesirable microorganisms shall be processed to and maintained at a safe moisture level by:
   - Monitoring the aw of food;
   - Controlling the soluble solids / water ratio in finished food;
   - Protecting finished food from moisture pickup, by use of a moisture barrier, or by other means, so that the Aw of the food does not increase to an unsafe level;

n. Food such as, but not limited to, acid and acidified food, that relies principally on the control of pH for preventing the growth of undesirable microorganisms shall be monitored and maintained at a pH of 4.6 or below by: - Monitoring the pH of raw materials, food in process, and finished food; - Controlling the amount of acid or acidified food added to low-acid food;

o. If ice is used and comes in contact with food products, it must be made from potable water and be in a sanitary condition;

p. Areas and equipment that are used to process human food should not be used to process non-human food-grade animal feed, or inedible products unless there is no possibility of contaminating the human food;

q. A coding system should be utilized that will allow positive lot identification in the event it is necessary to identify and segregate lots of food that may be contaminated.
   - Records should be kept for a period of time that exceeds the self life of the product, except that
   - Records need not be kept beyond two years.

Hazard Analysis and Critical Control Points (HACCP)

Fruit and vegetables as raw materials start as living cells and as such can vary in composition, colour, flavour structure and nutrient content. Thus a key part of describing a process is the preparation of detailed specifications for ingredients and packaging materials to ensure that final product performance and composition specifications can be met with the specified process and equipment. This is only possible if ingredients are preprocessed to the desired specifications. Thus the processing of foods must be separated into two broad areas: preparation of raw fruit and vegetables for further processing to consumer products or ingredients; assembly of preprocessed ingredients to finished consumer products.

The assembly of preprocessed ingredients to finished consumer products will be discussed to illustrate principles used to describe a process for quality control. These principles apply to the conversion of raw foods to processed products. The chief differences are the variations in raw product specifications and certain washing, peeling, size reduction and blanching or heat treatment steps needed to convert highly variable raw materials to standardized ingredients.

Typical operations using in converting raw fruits and vegetables to processed- ingredients for packaging and preservation or prior to a further heat treatment processing are as follows:

Process Description for Quality Control

One of the most important specifications of a product is its safety in terms of microbial contamination, freedom from hazardous chemicals and absence of foreign materials such as metal pieces, non-edible parts such as pits or woody stem material and dirt, insect parts or other extraneous material.

The microbial safety of processed fruits and vegetables is of prime importance from a quality viewpoint. The following analysis is to provide a description of a process to help ensure that microbial safety can be achieved with a minimum opportunity for failure of the finished product to meet specifications.

Process Operations

Food processing steps require a detailed description when microbial safety is a concern. The reason for this is that ingredients, packages, equipment, and the surroundings all potentially can contaminate the final product with pathogenic or spoilage microorganisms.

Further, if the conditions of pH (acidity), temperature, moisture and nutrient level are suitable, rapid microbial growth is possible on processing equipment and in the food itself. Thus while microbiological specifications can be written for incoming ingredients, actual microbial counts can increase during each process step if the process is not designed properly.

One of the first requirements for the description of a process is to determine if an individual process step will increase (+), decrease (-) or result in nonchange (=) in the microbial content of food undergoing the processing step. This can be determined from the chemical and physical conditions of the food and surroundings in each processing step.

The physical and chemical conditions of a food passing through a processing step can be recorded for each process step needed to prepare a finished product.

Individual processing steps are identified in the preparation of the canned juice drink from warehouse inventory to just prior to can labelling. Processing steps needed to prepare the package for filling and sealing are listed on the right side of the chart.

For purpose of analyses the juice drink must have a pH of 3.8, a fill temperature of no lower than 190° F and must leave the cooling tunnel at a temperature below 90° F. These requirements are highlighted in their appropriate column to ensure that the proper process control and quality control procedures are in place.

Equipment sanitation, prevention of air borne contamination during filling, seal integrity and rapid cooling are essential parts of the process.

Even with these requirements in control, the heat treatment given the products (heating to 200° F), with a variable 0-60 min hold) is not sufficient to inactivate Clostridium botulinum. Thus the product must be cooled to 35° F within 40 minutes to prevent any possible germination and outgrowth of Clostridium botulinum spores.

Further, the manufacturer of this product must provide a means to ensure that the product is used within a specified time and is always kept at or below 35° F if it is to be released to the public.

References

Chan, H.T. Jr. 1984. *Handbook of Tropical Foods*. New York/Basel: Marcel Dekker.

FAO. 1991. Post-harvest and processing technologies of African staple foods: a technical compendium. *Agricultural Services Bulletin* No. 89. Rome: FAO.

Harris, R.S. and Karmas, E. 1975. *Nutritional Evaluation of Food Processing*. 2nd Edition.AVI Publishing Co.,Westport,Conn.

Potter, N.N. 1984. *Food Science*. 3rd Edition. AVI Publishing Co. Westport, Conn.

Van Arsdel, W.B. and Copley, M.J. 1964. Food Dehydration. Vol. II *Products and Technology*. AVI Publishing Co., Westport, Conn.

Bibliography

Abers, J.E., Wrolstad, R.E. 1979. Causative factors of colour deterioration in strawberry preserves during processing and storage. *Journal of Food Science*. 44, 75–78.

Amoriggi, G. 1990. *Technical Report on Small-Scale Processing of Agricultural Products*. FAO. Strengthening of Agricultural Marketing Service GUY/86/003. Rome, Italy. 23 p.

Anon. 1983. Jams and jellies. *Institutional Distribution*. 19(12), 218, 222–224.

Ashurst, P.R. (Editor). 1995. *Production and Packaging of Non-Carbonated Fruit Juices and Fruit Beverages*. Blackie Academic and Professional.

Bachmann, J. & Earles, R. 2000. *Post-harvest handling of fruits and vegetables*. ATTRA. 24 pp.

Baker, R.A., Berry, N., Hup, Y.H. 1996. *Fruit preserves and jams in processing fruits—Science and Technology,* Vol 1. Biology, Principles and Application. Technomic Publishing Co., Inc.

Beaulieu, J.C. & Gorny, J.L. 2001. *Fresh-cut Fruits.USDA Handbook*. 66: 16-30.

Brett, A et al. 1996. *Producing Solar Dried Fnlits and Vegetables jor Micro- and Small-scale Rural Enterprise Development: A Series of Practical Guides*. Chatham, UK: NRI

Central Food Technology Research Institute, 1977, *Home Scale Processing and Preservation of Fruits and Vegetables*,Mysore, India.

Chen, C.S., Shaw, P.E. and Parish, M.E. 1993. Chapter 5 *In* Nagy, S., C.S. Chen, & P.E. Shaw, eds. *Fruit Juice Processing Technology*. AgScience, Inc., Auburndale, Florida.

Colin, D. 1992. Recent trends in fruit and vegetable processing. In *Food Science and Technology Today*, 7, (2)), pp. 111 - 116.

DeLong, D. 1979. *How to Dry Foods*. Los Angeles: HP Books.

Desrosier, N.W. and Desrosier, J.N. 1977. *Technology of Food Preservation*, 4th Edition, AVI Publishing Co. Westport, Conn.

Downing, D.L. 1972. *Homemade Fruit Juice Press*. Special Report No. 8. N.Y. State Agric. Expt. Sta., Geneva, NY.

FAO. 1988. *Rural Processing and Preserving Techniques for Fruits and Vegetables.* Rome, Italy. 59 p.

Feinberg, B. 1973. "Vegetables". In *Food Dehydration*, 2nd Edition, Vol. 2. W.B. Van Arsdel, M.J. Copley and A.I. Morgan, Jr.(Editors). AVI Publishing Co., Westport, Conn.

Fellows, P. and Hampton, Ann. 1992. *Small-scale Food Processing. A guide to appropriate equipment.* Intermediate Technology Publications. London, UK.

Fellows, P.J., 1993, *Food Processing Technology*, Woodhead Publishing, Cambridge, UK

Giridhari Lal, Siddappa, G.S., Tandon, G.L. 1986. *Preservation of fruits and vegetables.* Indian Council of Agricultural Research. New Delhi.

Harris, R.S. and Karmas, E. 1975. *Nutritional Evaluation of Food Processing.* 2nd Edition.AVI Publishing Co.,Westport,Conn.

Hobson, P. 1994. *Making and Using Dried Foods,* Pownal, VT: Storey Communications, Inc.

Holdsworth,: S.D. 1983. *The Preservation of Fruit and Vegetable Food Products.* MacMillan Press, London. 159 p.

Ihekoronye, A.I., and Ngoddy, P.O., 1985, *Integrated Food Science and Technology for the Tropics,* Macmillan Press Ltd., London, UK.

Kim, J.G. 2007. Fresh-cut market potential and challenges in Far-East Asia. Acta. Hort. 746, 53-60.

Kratz. 1993. Jam, jellies, marmalade II. The phenomenon of syneresis and method of manufacturing. *Food Marketing & Technology* 17–21.

MacDonald, I. and Low, J., 1984, *Fruit and Vegetables*, IT Publications, London, UK.

Mc Williams, Margaret and Paine, Harriet. 1977. *Modern Food Preservation.* Plycon Press, California, USA. 198 p.

Miller, M. et al. 1981. *Drying Foods at Home.* University of California, Division of Agricultural Science, Leaflet 2785.

Nagy, S., Chen, C.S. and P.E. Shaw. 1993. *Fruit Juice Processing Technology.* AgScience, Inc., Auburndale, Florida.

Potter, N.N. 1984. *Food Science.* 3rd Edition. AVI Publishing Co. Westport, Conn.

Siripanich, J. 2000. Post-harvest physiology of tropical fruits. *Acta Horticulturae,* 575.

Tindall, H.D., 1983, *Vegetables in the Tropics*, MacMillan Press Ltd., London.

Tressler, D.K. and Joslyn, M.A. (Editors). 1961. *Fruit and Vegetable Juice Processing Technology.* 1st edition. AVI Publishing Co. Westport, CT.

UNIFEM. 1988. Fruit and Vegetable Processing. *Food Cycle Technology Source Book* No. 2. Photosystem SRL, Rome, Italy. 67 p.

Van Arsdel, W.B. and Copley, M.J. 1964. Food Dehydration. Vol. II *Products and Technology.* AVI Publishing Co., Westport, Conn.

Watada, A.E. & Qi, L. 1999. Quality of fresh-cut produce. *Post-harvest Biology and Technology,* 15(3): 201-205.